P9-DXH-412

The Cyber Effect

The Cyber Effect

A Pioneering Cyberpsychologist Explains How Human Behavior Changes Online

MARY AIKEN, PhD

SPIEGEL & GRAU
NEW YORK

This is a work of nonfiction. Some names and identifying details have been changed.

Published in the United States by Spiegel & Grau, an imprint of Random House, a division of Penguin Random House LLC, New York.

SPIEGEL & GRAU and the HOUSE colophon are registered trademarks of Penguin Random House LLC.

Library of Congress Cataloging-in-Publication Data
Names: Aiken, Mary, author.
Title: The cyber effect : a pioneering cyberpsychologist explains how human behavior changes online / Mary Aiken.
Description: New York : Spiegel & Grau, 2016. | Includes index.
Identifiers: LCCN 2016007455 | ISBN 9780812997859 (hardback) | ISBN 9780812997866 (ebook)
Subjects: LCSH: Human behavior. | Internet users—Psychology. | Interpersonal relations—Psychological aspects. | Social psychology. | BISAC: PSYCHOLOGY / Social Psychology. | PSYCHOLOGY / Movements / Behaviorism. | TECHNOLOGY & ENGINEERING / Social Aspects.
Classification: LCC BF199.A37 2016 | DDC 155.9—dc23 LC record available at http://lccn.loc.gov/2016007455

Printed in the United States of America on acid-free paper

randomhousebooks.com
spiegelandgrau.com

2 4 6 8 9 7 5 3 1

First Edition

Book design by Caroline Cunningham

For P. L. K. & J.

Children are the world's most valuable resource and its best hope for the future.

—John F. Kennedy

CONTENTS

The Cyber Effect

PROLOGUE

When Humans and Technology Collide

I am sitting on a hard, cold bench. My back is against a concrete wall in a police briefing room somewhere in South Los Angeles—in a neighborhood known for gangs, crime, poverty, urban decay, and, twenty years ago, brutal race riots. It is 4:45 in the morning. I haven't eaten anything for hours. Not a wise move. My stomach is churning, a combination of hunger, jet lag, and apprehension.

LAPD lieutenant Andrea Grossman begins the police briefing—and explains how, in an hour or so, a special task force will be arresting the biggest human trafficker in the United States and one of California's "Most Wanted." About forty law-enforcement officers will be involved in the operation, a team of experienced professionals pulled from the FBI, Homeland Security, Internet Crimes Against Children (ICAC), the California State Police, and the LAPD. And then there's me, the one person at the briefing without a gun. Only sworn officers are allowed to carry them.

Back in Ireland, where I'm from, it is rainy. The spring drags on, gray and wet. I think about my cozy office in Dublin, my library, my desktop computer, and my quiet academic life. Except my life is not so quiet lately. Over the past decade, while establishing myself as a forensic cyberpsychologist, I have traveled the world to meet with other experts in my field, conducted research, worked with law enforcement, attended conferences, and given hundreds of talks, seminars, workshops, and presentations. The field of cyberpsychology is new and still emerging, and each year it draws more interest. The sense of urgency is escalating. I think most of us who work on the front lines can feel it, along with a profound sense of loss of control. Our lives are changing, and human behavior is evolving. As a cyberbehavioral scientist, I believe this is because people behave differently when they are interacting with technology than they do in the face-to-face real world.

Some changes have occurred so quickly that it has become difficult to tell the difference between passing trends, still evolving behavior, and something that's already become an acceptable social norm. In this book, to make things simpler, I will be referring to face-to-face reality as "real life" or the "real world" to set it apart from cyberspace, although I am fully aware that what happens there can be as real as anything. New norms created online can migrate to real life. So what happens in the virtual world affects the real world, and vice versa.

Whenever I am asked to talk about my work, I start off with the formal definition. Cyberpsychology is "the study of the impact of emerging technology on human behavior." It's not just a case of being online or offline; *cyber* refers to anything digital, anything tech—from Bluetooth to driverless cars. That means I study human interactions with technology and digital media, mobile and networked devices, gaming, virtual reality, artificial intelligence (A.I.), intelligence amplification (I.A.)—anything from cellphones to cyborgs. But mostly I concentrate on Internet psychology. If something qualifies as "technology" and has the potential to impact or change human behavior, I want to look at how—and consider why.

Time is not on my side. My work is always in a race with technology. This presents a major challenge to how academics normally study a phenomenon. As scientists, how can we possibly keep pace with the tech changes we are seeing in our lives, in our behavior, and in our society? A good longitudinal study, which looks at human behavior over time and allows a researcher to make conclusive scientific statements, can take anywhere from a couple of years to a few decades. That's several lifetimes in tech-terms. And given what I've seen already, particularly the new norms that are rapidly being created due to an accelerated form of socialization that I call *cyber-socialization,* I don't think we should sit around waiting for answers.

The good news: Some aspects of Internet psychology have been studied since the 1990s and are well known and documented. The effect of anonymity online—or perceived anonymity—is one example. It's the modern-day equivalent of that superhero power invisibility. The subject of some fascinating studies across many disciplines, anonymity has an impact that cannot be underestimated. It also fuels *online disinhibition,* another important contributor to other effects. I have been involved in a dozen different research silos, and have studied everything from organized cybercrime to cyberchondria, health-anxiety facilitated and amplified by doing online medical searches, and the one thing I have observed over and over again is that human behavior is often amplified and accelerated online by what I believe to be an almost predictable mathematical multiplier, the *cyber effect,* the $E = mc^2$ of this century.

Altruism, for example, is amplified online—which means that people can be more generous and giving in cyberspace than they are face-to-face. We see this phenomenon in the extraordinary growth of nonprofit crowdfunding online. Another known effect of cyberspace is that people can be more trusting of others they encounter online, and can disclose information more quickly. This leads to faster friendships and quicker intimacy, but it also means that people tend to feel safe when they aren't. Due to *online disinhibition effect* (ODE), individuals can be bolder, less inhibited, and judgment-impaired. Almost as if they were drunk. And in this less-inhibited state, like-minded people can find one another instantly and easily, under a cloak of anonymity, which results in another effect: *online syndication.* I will explain these cyber constructs and "effects" in detail in this book, and they are described in a glossary of terms, but ultimately they will be fully understood and evaluated only by empirical science—by undertaking intensive experimental studies, manipulating variables, and identifying cause and effect. But cyber isn't a lab with white mice and levers. We are talking about a complex matrix of human data that is manifested in a virtual context. It involves painstaking digital forensic and cyberbehavioral detail.

There is an expression, "God is in the details," that resonates with my work. *Forensic science* is the study of the physical evidence at a crime scene—fibers or bodily fluids or fingerprints. In TV terms, think *CSI*. *Forensic psychology* is the study of the behavioral evidence left behind at the crime scene, what we like to call "the blood spatter of the mind." Then there's my area, *forensic cyberpsychology,* which focuses on the cyberbehavioral evidence of a crime scene, or, as I like to think of it, the *cyber footprint.* It was the great forensics pioneer Edmond Locard, sometimes called "the Sherlock Holmes of France," whose exchange principle put forth the basic premise of forensic science: "Every contact leaves a trace." (Your fingerprints are now all over this book.) This is just as true in cyberspace. Almost everything we do online generates digital exhaust, digital dust, and digital prints. This digital evidence can help law enforcement investigate criminal behavior, whether the crimes take place in cyberspace, across the world, or down the street.

It was the pursuit of that kind of data that led me to Los Angeles. I was conducting a study with the Specialists Group at INTERPOL, the world's largest international police organization, about youth risk-taking online and, hoping to accumulate data, I got in touch with Lt. Grossman at the LAPD. We had met previously at a conference at the INTERPOL headquarters in Lyon, France. I'd been impressed by Lt. Grossman and her work in the field of cybercrime. When she agreed to see me and discuss the INTERPOL project, I flew to California to meet her team.

Police can be very skeptical about academics descending from their ivory tower who are hungry for data but have very little understanding of the nitty-gritty nature of frontline law enforcement. So I was pleased that Lt. Grossman asked if I'd be interested in getting some work experience with the LAPD.

"Of course," I replied, assuming that she was talking about a type of internship at her police precinct, where I would sit in on meetings, but she had something a little more proactive in mind.

"How would you like to suit up and come on an operation?" she asked, going on to explain that the identity and location of a trafficker of child abuse images, videos, and other materials had been determined by using cyber-forensics. Lt. Grossman thought this would interest me, as an academic observer.

"Uh . . . yes," I stammered. "You mean, suit up, like S.W.A.T.? When?"

"Tonight."

My work involves the scientific investigation of behavior online—from the prediction of developing behavior, such as *cyber juvenile delinquency* (hacking), to profiling typologies for evolutions of criminal behavior (cyberstalking). I explore machine intelligence solutions to big-data problems (such as technology-facilitated human trafficking) and *intelligence amplification* (I.A.) solutions to child-related online sex offending. This is all demanding work that I have been trained to do and have learned to handle. But real-world frontline police work? S.W.A.T. takedowns? I have very little experience of that.

In my hotel room, later that evening, I dressed in black—the forgettable, blend-into-the-woodwork uniform of forensic experts worldwide. (Why hemorrhage data at fifty paces by wearing a pale-pink blouse to demonstrate that you're feeling vulnerable, a splash of yellow for optimism, or a pattern to make you appear interesting?) Then, at 3:30 a.m., I grabbed a bottle of water, went downstairs to the lobby, and told the reception desk that a group of LAPD officers would be coming soon to pick me up.

The concierge looked at me skeptically.

"I've done nothing wrong," I assured him. "I've been asked to observe a mission. That's all."

That's how I wound up here, before dawn, in an LAPD briefing room. The weather in L.A. is always agreeable, so they tell me, but it is unexpectedly chilly this morning. Fortunately I have a bulletproof vest and a steel ballistic helmet to keep me warm.

"Resistance is always expected," the briefing officer says. "If an officer goes down, step over them. Just keep moving forward. If you go down, stay down."

I glance at the briefing book in my lap. It includes directions to the nearest hospital. *If you go down, stay down. . . .*

Faced with uncertainty—and potential danger—I adopt an attitude that has served me well in life: Hope for the best, expect the worst. And that turns out to be a pretty good motto for almost any endeavor, whether you are living in the real world or online. Each time we join a new social network, download an app, pay a bill online, buy our children a new digital device, or meet someone on a cyber-dating site, we are faced with a steep cyber learning curve and can quickly encounter new challenges and risks. Hiking up a sheer mountain trail to enjoy a breathtaking view is one thing. Jumping off the summit to paraglide down is another. Some risks are worth taking. Others are just unnecessary. Which is which? That is what this book is about.

"Let's roll!" Lt. Grossman calls out. Twenty chairs slide back at once. Boots stomp. Guns clank. I reach for my helmet and pause for a second and think, not for the first time that morning, *How on earth did I get here?*

Where Am I?

We are living through a unique period of human history, an intense period of flux, change, and disruption that may never be repeated. A seismic shift in living and thinking is taking place due to the rapid and pervasive introduction of new technologies to daily life, which has changed the way we communicate, work, shop, socialize, and do almost everything else. This moment in time is not unlike the Enlightenment (1650–1800), when there were also great shifts in awareness, knowledge, and technology, accompanied by great societal changes.

Enlightenment delivers new freedoms. And the new freedoms allowed online are heady, thrilling, and enticing to billions of people. The concept of absolute freedom is central to the ideology of the Internet. But can this freedom corrupt? And can absolute freedom corrupt absolutely? More freedom for the individual means less control for society.

Some changes have been seductive and incremental—and have caused psychological norms to creep into new places. You barely noticed until, one day, suddenly you see a baby in a stroller being handed an expensive smartphone to play with or you see a toddler expertly swipe a touchscreen with a chubby finger. Or maybe you walk into a shopping mall and notice a group of kids huddled together solemnly looking at their devices—and not one another. So near and yet so far!

Or something might have hit you closer to home, like an increasingly distant and uncomfortable feeling in your relationship or marriage because your partner is spending hours alone with his or her computer—chatting and cyber-flirting with new friends worldwide, bingeing on Netflix, consumed by online shopping, or obsessed by the plethora of pornography sites so readily available now online.

The Internet is omnipresent, always delivering rich, stimulating content—all day, all night, always on. Between the years 2000 and 2015, the number of people with access to the Internet increased almost sevenfold—from 6.5 percent to 43 percent of the global population. At the Davos Summit in January 2016, it was announced that more than 3.2 billion people are now online. In less than ten years the number of cellphone subscriptions has grown from a little more than 2 billion in 2005 to more than 7 billion in 2015. The number of hours people spend on mobile phones is escalating rapidly each year, jumping an average of 65 percent in a two-year period. The same study found that mobile phone users checked their devices more than fifteen hundred times a week, and there are several apps that will count that for you, if you need a little help managing your habit.

The number of minutes per day that you spend checking your phone and scrolling through social media posts is not insignificant. To a researcher like me, who studies human behavior at the minute-by-minute level—in digital dust and footprints—these minutes indicate how a person is living—what they do and don't do. This is called *pattern of life analysis,* or how people live online. In the home, these minutes are not spent doing other things—reading a book to a child, playing with a toddler on the floor, chatting with your family at the dinner table, talking with your partner before bed. When you are checking your phone or spending time surfing websites, you are effectively in a different environment. You have gone somewhere else. You are not present in real-world terms.

Let me raise a question, one that has been fiercely debated by technologists: *Is cyberspace an actual place?*

My answer is unequivocal: Yes, it is. Cyberspace is a distinct place. You may be accessing it from a familiar environment, like the comfort of your own home, but as soon as you go online, you have traveled to a different location in terms of your awareness or consciousness, your emotions, your responses, and your behavior—which will vary depending on your age, your physical and mental development, and your distinct set of personality traits.

Instinctively, we know this is true. Most of us have felt "lost" in cyberspace and realized—as if waking from a dream—that we've burned dinner, run late for an appointment, or forgotten to turn off the sprinklers. This is due to the fact that, in the real world, most people have learned to keep track of time effectively. Online, though, there's a *time-distortion effect.* (Try this the next time you log on: Turn off your clock display, and every so often test yourself to see how well you can estimate the passage of time.) As complex as human beings are, and as adaptable, psychologists know from a myriad of studies and research that when an individual moves to any new location—a new home, a new school, a new city, or a foreign country—his or her behavior will change or adjust. One's environment has a profound impact on one's physical bearings, something we know from work done in the field of environmental psychology, an interdisciplinary approach that looks at the interplay between individuals and their surroundings. And according to theories of development, an awareness of self comes through a gradual process of adaptation to one's environment. And as anyone knows who has moved or traveled, it can take time to absorb and acclimate to any new location or space. It can take a while to get your "sea legs," as sailors who shove off from land to live aboard a boat would say.

But many people deny the awareness that they've entered a new environment when they go online, so they remain ignorant—and are fooled by their sense that nothing has changed. They are sitting in their own homes, surrounded by familiar objects, after all, and their bodies are resting in the cushions of familiar chairs and sofas. In their minds, they have not "gone" anywhere. But the conditions and qualities of the online environment are different from real life. That is why our instincts, which were honed for the real world, fail us in cyberspace.

Naiveté and bad judgment about this environment can be evidenced every day—when we pick up a newspaper and see that a politician has distributed photographs of his genitals to horrified strangers, when a celebrity rants crazily on Twitter, or when another sex tape goes viral. Traditional authorities and support systems appear to be absent online—or they are just as confused as you are. As devices and gadgets change, and technology changes, the cyber environment changes with it, which impacts human behavior again. This causes upheaval to individuals, industry, finance, government, all of society. The more changes there are, the more new situations arise, creating only more confusion.

Psychologists know that living in a state of societal change is easier for some than others. But for most, trying to keep pace with recent technological changes has been dizzying. While many people are still finding their footing in this new environment, with all its new neighborhoods and new behaviors, there are many more changes yet to come. This can only result in more new situations and more confusion.

One sure way of coping with a state of constant flux is to become more knowledgeable about how the cyber environment affects all of us—how people, yourself included, may act there. Knowledge is power, and it's tremendously reassuring. A familiarity with the basics of cyberpsychology will help you answer the questions I hear all day, and would hear all night if I were to never sleep and just read my email.

Questions like:

- At what age can my baby start watching digital screens?
- Is it okay for a toddler to play with an iPad?
- Is there a connection between online gaming and ADHD in young boys?
- Should I allow teenagers to spend hours in the bathroom with their smartphones?
- Does technology contribute to social isolation?
- Can real relationships be formed in cyberspace?
- Why do people troll online?
- Should I be afraid of "the Deep Web"?

Cyber is not just a transactional medium, for things like passively viewing television or making a phone call. It is a highly interactive, highly engaging, and highly immersive environment—uniquely compelling and attractive to humans. Perhaps too compelling. What about your toddler who throws a tantrum when you ask for your tablet back, or your teenager who screams when Wi-Fi slows down, or your aunt and uncle who seem to be in a constant state of *tech rage* ("The computer's broken!"), or the fact that your grandmother on Facebook has made lots of new online "pen pals" in Nigeria?

Cyberspace is full of place names—social networks, forums, sites—and once there, we join up with a far larger group than we've been with before, which also makes this environment distinct. There are now billions of people online. This has prompted a lot of new situations and confusion. With such a wide array of new friends and contacts in your life, it's crucial to know more about human behavior—and understand how it changes online. Our instincts have evolved to handle face-to-face interactions, but once we go into cyberspace, these instincts fail us. We are impaired, as if we had been given keys to a car but not learned how to drive. We need more tools and more knowledge. Because if you spend time online, you are likely to encounter a far greater variety of human behavior than you have before—from the vulnerable to the criminal, from the gleeful and altruistic to the dark and murderous.

My focus on cyber-forensics in my work with law enforcement means that I witness both the best and the worst aspects of human behavior manifested online. I like to say that technology was designed to be rewarding, engaging, and seductive for so-called normal populations. But did anyone really think about how it would impact abnormal, deviant, criminal, and vulnerable populations?

Considering those risks is part of my work too.

How to Read This Book

We all know about the incredible benefits of the Internet. I could talk all day about them—the convenience, connectedness, affordability, creativity, altruism, educational and commercial opportunities, entrepreneurship, and cultural exchange. I'm pretty sure you are aware of these things too. An army of marketing experts working for all the biggest tech companies and conglomerates do nothing but dream up new and better irresistible products and new and better ways to sell them to us.

They are supergood at convincing us of the necessary features of these gadgets and software and apps and touchscreens.

My job isn't to criticize technology. Good science focuses on balance. If I seem to focus on many of the negative aspects of technology, it is in order to bring the debate back to the balanced center rather than have one driven by utopian idealism or commercialism. My job is just to provide the best wisdom possible, based on what we know about human beings and how their cognitive, behavioral, physiological, social, developmental, affective, and motivational capabilities have been exploited or compromised or changed by the design of these products.

Technology is not good or bad in its own right. It is neutral and simply mediates behavior—which means it can be used well or poorly by humankind. This understanding is fundamental to my work. This is no different from how we regard automobiles and drunk driving. Any technology can be misused.

One of my earliest influences was J.C.R. Licklider, an American psychologist and computer scientist who in 1960 wrote a seminal paper, "Man-Computer Symbiosis," which predated the Internet but foretold the potential for a symbiotic relationship between man and machine; in fact, you could say he was the first cyberpsychologist. I read "Lick" with amazement at his ability to gaze into the future with such clarity and wisdom. Early on I was also drawn to the work of Patricia Wallace, who wrote *The Psychology of the Internet,* an influential academic book and popular success in 1999. Soon afterward, I became aware of John Suler, a clinical psychologist and pioneer, the acknowledged "father of cyberpsychology," who has been working in this area since the nineties and wrote *The Psychology of Cyberspace,* first published as a digital book in 1996. John has really captured the essence of cyber in his work and has explored the potential benefits and hazards of cyberspace and characterized the way people tend to behave in the online environment.

Just as I was embarking on my own study and research, I reached out in cyberspace to John. This led to an exchange of emails, which led to an eventual face-to-face meeting at Rider University, in New Jersey, his academic home. They say it's hard to meet your idols. But in my case, I just wish I'd worn the right shoes.

It was a grueling hot day, and John had just come from a lecture when I arrived on campus. He wanted to stretch his legs a bit. "Let's walk while we talk," he said. Then, with the air of a Socratic philosopher striding across the Acropolis, he set off at warp speed across the quad. John is a tall man, and each stride meant about four hurried steps for me. To prepare for our meeting, I had carefully considered various cyberpsychological constructs that I thought we might discuss, but I didn't think it would happen outdoors in blistering heat, or while I was wearing heels that were unsuitable for uneven terrain, much less a forced-march pace that would make a marine weep. In many ways, all the rest of us are still trying to keep up with John.

Over the past decade, he has become my great friend and colleague. Some of his groundbreaking constructs and observations inform a number of the concepts addressed in this book. In recent years, I have had the pleasure of meeting a growing group of like-minded researchers worldwide, who share ideas with me and collaborate on studies. I am thrilled to showcase an impressive body of work in the chapters to come. Approximately thirty peer-reviewed journals now publish an estimated one thousand articles every year on topics related to cyberbehavior, a field that is expected to enjoy exponential growth in the next decades due to the pervasive and profound impact of technology on humans.

Like other fields of scientific endeavor, mine is a land of jargon and caution. The behavioral sciences have been blindsided by developments in technology to a certain extent. In the late 1990s colleagues of mine referred to the Internet as a passing phenomenon. In the mid-2000s they said that people would never use online social media platforms to communicate. Now fifteen years and billions of people later . . . a game of catch-up is going on.

Academics are great at finding complicated ways to not really say what we mean. Our academic papers are littered with hedging adverbs like *arguably, plausibly,* and *questionably.* We seem to enjoy adding an *-ably* to as many words as possible, hoping to render our sentences harmless. Some researchers employ what I call "sleight of word" as a career-protection mechanism, just in case, at some point in time, an idea might be proven wrong. But I don't believe scientific breakthroughs are achieved by metaphorically sitting on the fence. On the cyber frontier, we need scientists who are prepared to nail their colors to the mast and back their own informed instincts. Of course we need evidence-based studies over time, but how long can we wait?

Babies are being born, kids are growing up, and lives are being changed. Society is being reshaped. We need to talk about this now.

In hopes of reaching a wider audience, I have tried to make this book as practical and straightforward as I know how. I have tried to make the science comprehensible and spare you too many stats and studies. For those who share this affinity or are interested in a deeper dive, there are extensive chapter notes to draw on in the back of the book. They are written with a broad audience in mind as well.

To keep up with changing technology, and changing human behavior, my work requires creativity, flexibility, and an ability to juggle a lot of theoretical constructs. It's probably a good thing, then, that I haven't got the sort of brain that thinks in a linear way. I feel more like a depository of organized chaos, but this helps me identify patterns quickly and make intuitive leaps. My approach is transdisciplinary by necessity—drawing on psychology, sociology, anthropology, computer science, criminology, and network science. It can cross other academic boundaries too. I find that the different disciplines help to illuminate problems that are arising, and help to illuminate solutions too.

In the absence of longitudinal studies, I employ logic—a mixture of common sense and reasoning—then construct plausible arguments based on a body of knowledge and current observable phenomena and reports, which I hope will start some meaningful debates about human behavior online, something I feel is much needed. I have also drawn on those very special and uniquely human skills: insight and intuition. As the great robotic scientist Masahiro Mori said, "Do not ignore the small things." In science we should not be afraid to listen to ourselves or to pay attention to the little things. Mori himself was not reluctant to share his thoughts and suspicions—about humans and machines, about artificial intelligence, and about the need to take pleasure, even delight, in our intuitions. His approach inspires me. Academics need to reconsider how we handle behavioral problems that are evolving at the speed of technology. We need academic first responders.

Quite often, I have leaned on the investigative journalism of publications such as *Wired,* the *Washington Post,* the *New York Times,* and other reliable sources and legacy media to read emerging frontline reports of anecdotal evidence, see patterns of behavior, and try to make sense of them. In a field as rapidly evolving as the Internet, and the technology using it, we need good journalism more than ever.

In the nine chapters to follow, I have arranged material into areas of special concern—as well as my own focus. The impact of technology on human behavior begins at birth and ends at death, so I have chapters that deal with all age groups—from babies, toddlers, kids, and teens to adults. In chapters about addiction and compulsive behavior, I've looked at ways that some types of problematic behavior can be enormously impacted by the online environment. And in a chapter about the phenomenon of cyberchondria, I've argued that the prevalence of the online medical search has resulted in a rise in unnecessary doctor visits and risky surgical procedures.

The frightening revelations in this book, and the chapter on the Deep Web, are not included simply for kicks and thrills. The dark hidden corners of the Internet where criminals syndicate and a black market is thriving are things every single person online should know about. Why? Because more and more young people are being enticed to go there, driven by a combination of adolescent risk-taking and curiosity. Somehow they've gotten the wrong impression that it's perfectly safe in the Deep Web, even fun. But it isn't.

My own particular concern is the impact of technology on the developing child. The Internet has opened the world up to our children, yet it gives the world access to them too. I don't think most people know enough about this. There is a great paper in the journal *Pediatrics* on the impact of technology on the developing child entitled "The Good, the Bad, and the Unknown." It's that last part, *the unknown,* that really bothers me. As the clinical psychologist Michael Seto has said, "We are living through the largest unregulated social experiment of all time—a generation of youth who have been exposed to extreme content online."

What will happen to this generation over time? What is the impact in terms of exposure to the harsher and bleaker aspects of the Internet?

CSI: Cyber

The raid on the house in South Central L.A. was as terrifying as you might imagine, and I have to confess that as our convoy pulled up outside the target house, I turned to Lt. Grossman to ask if I could stay in the armored police car rather than move forward with the unit.

"No, Mary, it's not safe," she replied.

The armored car wasn't safe? Wow. I thought, *What am I getting into?* The next twenty minutes went by in a blur. There was a lot of shouting, banging doors, barking orders with guns drawn, handcuffings, and arrests. As an observer, it was both frightening and fascinating. I stood in the background, next to a wall of the living room where the suspect was apprehended, and I found myself tapping the wall, hoping it was solid concrete and would protect me from stray bullets. No bullets were fired, I am happy to report. The raid was a complete success, the kind of slick professional operation that this LAPD unit carries out several times a week. The main suspect was taken immediately to a mobile on-site LAPD computer forensics field truck, known as "the beast," where he confessed.

Once it was all over, the police team relaxed and tucked into a hearty feast of breakfast burritos while I sat quietly, sipping bottled water in a state of relief and shock, still swaddled in my protective gear. I have been asked to go out with Lt. Grossman and her team a few times since, but I assured them that my real-world frontline policing experience is truly complete. I have the utmost respect for the work carried out by first responders in law enforcement—day in, day out—but participating as an observer in an exercise like this served to reinforce this respect. And the truth is, I don't think that I am cut out for frontline active service in the real world—but I am happy to serve on the cyber frontier.

Besides, my real job is challenging enough—finding risks in places where we feel perfectly safe. Each year has brought more studies in my field and more discoveries. While conducting my research, I have had a chance to meet and speak with leaders in law enforcement and policy makers in government around the world, and have engaged as an academic with Europol, INTERPOL, the FBI, and the White House. In 2012, supported by a great mentor and colleague, Professor Ciaran O'Boyle, I founded the CyberPsychology Research Centre in Dublin, now an international network designed to support and nurture cutting-edge research projects, and most recently found myself spending a good bit of time in Hollywood, working on the television show *CSI: Cyber,* inspired by my work. In the show, Patricia Arquette plays Avery Ryan, a special agent in the FBI Cyber Crime unit who is tasked with solving high-octane crimes that "start in the mind, live online, and play out into the real world." That describes my work perfectly.

Factoring the Human

Earlier in this prologue I asserted my view that the Internet is distinct from the so-called real world, but that I don't mean to suggest that what happens there isn't real. And in terms of human behavior, what happens online is a little like one of those evolving flu viruses or Ebola. Once behavior mutates in cyberspace, where a significant number of people participate, it can double back around and become a norm in everyday life, something I call *cyber-migration.* This means that the implications of the online experience and environment are ever evolving and profound, and impact us all—no matter where we live or spend time.

When I studied psychology as an undergraduate, we used to say that the problem with the field was that for too long it had "lived on a diet of white mice and college student surveys." Something similar can be said of technology: For too long it has lived on a diet of data, devices, and tech experts. Now it's time to turn our focus to the greater socio-technological implications. How have these advancements changed human behavior and society? It is time to consider that awkward entity, *Homo sapiens,* whose thumbs are too big for cellphone keypads, whose bodies are too clumsily shaped for wearable technology design, and whose memory is too weak to retain multiple ten-digit passwords. In other words, it's time to factor in the human. Sometimes our excitement about technology has prevented us from seeing the bigger picture.

In the midst of the human migration to cyberspace, it is important to examine what's behind us, where we are now, and what lies ahead. Like travelers heading off on an adventure, we need to be careful not to rush too quickly out the door without making sure we are carrying things we need for the journey we're undertaking. There are some important things, aspects of human life that have served us well for centuries and are crucial to our survival, that we can't afford to lose or abandon on this journey. This is where the discipline of cyberpsychology can be invaluable, delivering insight at the intersection of humans and technology. My hope for this book is to do just that.

CHAPTER 1

The Normalization of a Fetish

Human behavior has always been affected and shaped by technology, but there has been no greater influence, as far as I can see, than the advent of the Internet. You don't have to be an expert in the subject of online behavior to have observed that something about cyberspace provokes people to be more adventurous.

The illusion is that the cyber environment is safer than real life—and connecting with other people online somehow carries fewer risks than face-to-face contact. But our instincts were trained and honed for the real world, and in the absence of real-world cues and other subtle pieces of information—facial expressions, body language, physical spaces—we aren't able to make fully informed decisions. And because we aren't face-to-face when we are communicating and interacting with others online, we can be anonymous or, more important, we feel we are. As discussed in the prologue to this book, we can feel freed up and emboldened online. People can lose their inhibitions and in a way "act drunk" because, for some, being in the cyber environment can impair judgment and increase impulsivity, somewhat similar to the way alcohol can. Disinhibition is facilitated by the environmental conditions of cyberspace—by the perceived lack of authority, the anonymity, as well as the sense of distance or physical remove.

You see this in the changing courtship rituals of self-curated *selfies* and *sexts,* and in flirtatious exchanges on social-networking sites. Online, we feel more comfortable being bold and explicit. An individual may demonstrate common sense, rationality, and restraint in real life and check these qualities at the door when entering cyberspace. Why?

In cyberpsychology, the explanation for this emboldened behavior is known as the *online disinhibition effect,* first introduced by John Suler and now highly cited and accepted by academics in the field. Another powerful factor comes into play, which I have studied and written about, *online escalation.* It is a construct or concept that I use to describe how problem behaviors become bigger—or amplified—online, as many of us have already witnessed in everything from supernegative exchanges via flaming emails, aggressive texts, and offensive posts to comment threads that are meant to provoke.

It isn't that technology is bad for us—or inherently negative. Problems occur when we are ignorant of its impact. Most people don't understand the effect that the environment of cyberspace can have on them. They think it's the same as anywhere else. Individuals who behave impulsively, or struggle with a tendency to act rashly, may be especially vulnerable. But due to the effect of online escalation, anyone can move more quickly into new behavior and new norms.

Later on, in a chapter about cyber romance, I will discuss the new ways people meet, make friends, form communities, and find meaningful personal connections online. In this chapter, though, I'll be exploring the impact of technology on a smaller slice of the population, specifically individuals with a fetish or paraphilia—what is considered atypical sexual behavior. Why bother looking so closely at one specific population online? By studying the extreme effects of technology on fringe or unusual behavior, we begin to see more clearly the implications of the cyber environment for all of us. As a forensic cyberpsychologist, I've seen this demonstrated time and again: Whenever technology comes in contact with an underlying predisposition, or tendency for a certain behavior, it can result in behavioral amplification or escalation.

I would argue that tendencies and vulnerabilities that cause the most distress in real life may become even more of a battle online. That goes for any behavior.

If these tendencies aren't destructive or risky, the impact can be fairly benign. If someone just loves to visit online forums for gardeners, there isn't much that's self-destructive about that. But there are many incidences of risky behavior becoming much riskier online, especially pathological and criminal behavior. Here's an example of what I mean: A stalker in the real world typically focuses on one victim at any given time, but a cyberstalker can stalk multiple victims simultaneously because technology makes that possible. Cyberstalking is considered an evolution of a real-world criminal behavior. Cyberspace is a breeding ground for mutations. Real-world behavior migrates there and escalates or accelerates. This can sometimes have serious implications in the real world.

A Case of Cranking

Jordan Haskins had grand dreams of making a difference in sparsely populated Saginaw County, Michigan, his hometown. The pale, clean-cut twenty-three-year-old man described himself as "pro-life, pro-family, pro-freedom, and pro-faith." In the summer of 2014, while still a student at Maranatha Baptist University, he announced he was running for political office, hoping to serve as a state representative to Michigan's 95th District. He posed for his campaign website photograph with the best smile he could muster, his image superimposed on a background of a billowing American flag, a sepia-toned Declaration of Independence, and artwork that appeared to be the three crosses of Calvary, synonymous with crucifixion, in dark outline.

The Republican candidate said his love of history, philosophy, religion, and politics had led him to run for office—and give back to society. He'd realized that his gift was for government.

"I've found my niche," he said, "my passion."

He had other passions, as it turns out. Before his campaign was fully under way, he found himself needing to explain his criminal record. He had been charged and pleaded guilty to four violations of trespassing on private and public property and unlawful use of government vehicles during a period of ten months in 2010 and 2011.

When interrogated by police at the time of his arrest, Haskins admitted that he had twice broken into a government yard where a mosquito-control pickup truck and sheriff's cruiser were parked in order to pull the spark-plug wires loose on the vehicles, sit in the front seat of the cars, and masturbate while listening to the engines sparking.

This behavior is called "cranking," he explained to police. He was sentenced to one year and eight months in prison in 2011. Parole was lifted just eleven days prior to the election. "I was in a messed-up state of mind mentally and emotionally when I did what I did," the candidate told the local media when asked in 2014 about his felony convictions. "That's the only way I can even explain it."

Cranking? What on earth possessed him to jump the fence into the police department parking lot, break into a car, and the rest . . .

"It was just the fun and the risk and the thrill," Haskins said.

And how had he discovered this bizarre fetishistic behavior?

"I read about it online."

Anatomy of a Fetish

At the back of New Age shops and hippie stores, next to the incense and patchouli oil, you'll find shelves of little carvings for sale. Inspired by the practices of Native American animism and West African religious cults, these small charms in the shape of animals are believed to be imbued with supernatural powers, carrying energy, communications from the spirit world—or simply good luck. Each of these objects is called a "fetish."

The use of the word *fetish* for an object that causes sexual arousal is attributed to French psychologist Alfred Binet, who is best known for devising the earliest intelligence tests. In common usage, a fetish describes a strong compulsion or desire for a particular condition or activity. A woman might say, lightheartedly, "I have a shoe fetish," meaning to suggest that she spends a disproportionate amount of her income on footwear. A man might say that stiletto heels excite him. In fact, feet and objects associated with feet are the most common fetishes. In psychology, *fetishistic disorder* is a mental health condition centering on the use of inanimate objects as a source of sexual satisfaction.

What causes one to develop a fetish in the first place? Sigmund Freud argued in his famous 1927 essay on fetishism that sexual fetishes were the result of arrested psychosexual development. Freud wrote, "When now I announce that the fetish is a substitute for the penis, I shall certainly create disappointment; so I hasten to add that it is not a substitute for any chance penis, but for a particular and quite special penis that had been extremely important in early childhood but had later been lost."

Interestingly, Freud believed that the predominance of foot and shoe fetishes could be explained by a child's early memories of being on the floor and peering up a woman's skirt.

Freud's views and theories have many critics. Numerous contemporary psychologists are skeptical of his therapeutic approach, psychoanalysis, and have distanced themselves from it. I happen to value the work of Freud because it often helps to illuminate complex behavioral issues, and these days psychoanalytic theory has also advanced well beyond Freud's ideas. More contemporary psychoanalytic approaches have different explanations for fetishes, and describe how they can become a way for individuals to maintain a cohesive sense of self. But in general, other schools within psychology now hold sway, from Jean Piaget's theory of cognitive development to Erik Erikson's theory of psychosocial development. Currently there's enormous interest in biopsychology, and how the brain and neurotransmitters influence our behavior.

I suspect if he were alive today, Freud would be keenly interested in the impact of technology on humankind. Since he argued that sexual impulses were at the root of most behavior, I'm sure he would identify sexual impulses as a contributor to some cyberbehavior.

Is sex behind all obscure behavior? Even Freud himself supposedly once said, "Sometimes a cigar is just a cigar."

Childhood experiences, though, do seem to explain some fetishes. As psychologist Robert Crooks and sex therapist Karla Baur write in their academic textbook, *Our Sexuality,* a boy may associate arousal with objects that belong to an emotionally significant person. The process of this association is sometimes referred to as *symbolic transfor-*

mation, when the object of the fetish becomes imaginatively endowed with the power or essence of its owner. In contemporary psychoanalysis the fetish is a *self object,* or an object that represents a significant other—usually a parent.

Some experts argue that fetishistic behavior has similarities with obsessive-compulsive disorder. And like OCD, once ingrained, a fetish is an urge that is difficult to deny—a behavior that is very hard to resist. The conditioning process is commonly thought by modern psychologists to reinforce the fetish. Like Russian physiologist Ivan Pavlov's famous dogs that salivated whenever they saw Pavlov's lab assistant, because they anticipated being fed, there is an anticipation of a reward when seeing the fetish object or thinking about the fetish behavior. In Haskins's cranking case, he had prior arrests as a juvenile for breaking into cars and joyriding, and perhaps these were thrilling adventures that he continually wanted to relive. Each time he broke into a car, his behavior was rewarded—reinforcing and strengthening the fetish. The more he took risks to do a little cranking, the more ingrained the behavior became. This is how conditioning works.

The growing number of websites catering to sexual fetishes involving cars (including pedalsupreme.com, which offers "pedal pumping, engine cranking, leg and foot videos and pictures") is a sign that this is an escalating trend. There is also a YouTube channel, pumpthatpedaldotcom, which offers a new installment every week. According to sexologist Dr. Susan Block, one possible psychological explanation for this behavior is that the " 'vroom' of the engine reminds them of their own libidos being revved up."

A fetish can be created in a lab. In one classical conditioning experiment a group of male subjects in a laboratory were repeatedly shown erotic slides of nude women. Occasionally, an image of women's boots would appear in the slideshow. Soon the men had a sexual response to the photograph of boots alone, and over time they responded to images of other types of women's shoes. If a photograph of a pair of women's boots could be used so easily to condition desire, imagine how the Internet with its infinite supply of images could inspire, instill, and sustain a fetish.

That brings us back to Jordan Haskins, who lost the election for the 95th District after he'd become a viral story—and the subject of unkind scrutiny and fascination. "You may not respect my policies, you may not respect my ideas, but you at least have to respect me as a person," Haskins said. But I don't think too many people did.

Politicians + Paraphilia

Unusual sexual behaviors were once labeled "deviant" or "perverted," even in clinical references. More recently, a less charged, less judgmental umbrella term, *paraphilia,* was adopted to describe a range of atypical behaviors that include fetishes. *Paraphilia* means "beyond usual or typical love." This alludes to the fact that these behaviors are not commonly associated with a traditional romantic relationship. Given the apparent rise in the number of websites catering to paraphilia, atypical sexual behavior may be becoming more common.

You don't have to look too far for examples in public life of individuals who, like Jordan Haskins, had everything to lose—and lost it—when unusual or bizarre sexual behavior was exposed. I'm thinking of U.S. congressman Anthony Weiner and the explicit photographs of himself that he sent to unsuspecting women he'd met online. I'm thinking of U.S. senator David Vitter and his urge to be diapered, which required the services of prostitutes. And I'm thinking of New York governor Eliot Spitzer, who insisted on keeping on his calf-length black socks when he had sex with a call girl. Remarkably enough, the marriages of these politicians, with the exception of Spitzer, survived after their unusual sexual habits came to light. The actor David Carradine, who was found dead and hanging inside a hotel room closet, wasn't so fortunate. According to an ex-wife, he was into an extremely dangerous autoerotic asphyxiation paraphilia called *hypoxyphilia,* consisting of practices that restrict the flow of oxygen to the brain for sexual arousal.

The *Diagnostic and Statistical Manual of Mental Disorders, Fifth Edition,* or DSM-5, the most recent version of the standard classification of mental disorders used by mental health professionals worldwide, breaks down paraphilia into eight main types of disorders:

exhibitionistic disorder, fetishistic disorder, frotteuristic disorder, pedophilic disorder, sexual masochism disorder, sexual sadism disorder, transvestic disorder, and voyeuristic disorder. The behavior being described in most of these terms is probably not too difficult to figure out, since the clinical words have entered common speech, with the exception of *frotteurism,* a disorder in which a person derives sexual pleasure or gratification by rubbing himself, or his genitalia, against another person—usually in a crowd (a good reason to avoid overly crowded subway cars at rush hour).

Of the eight paraphilia disorders, by far the most common are: *transvestic disorder,* which refers to an individual who experiences erotic or sexual gratification while cross-dressing; *exhibitionistic disorder,* the behavior engaged in by a "flasher" who shows his genitals to unsuspecting strangers; *voyeuristic disorder,* the clinical term for the activities of a "peeping Tom" who is gratified by spying on naked women; and finally, and most disturbingly, *pedophilic disorder,* describing the behavior of an individual who is sexually aroused by children.

Some of these disorders tend to cluster, meaning an individual could have more than one paraphilia. It's not uncommon to find fetishism, transvestism, and sadomasochism together.

Many dozens of other paraphilias have been written about—ranging from sexual attraction to amputees (*acrotomophilia*), dead people (*necrophilia*), pregnant women (*maiesiophilia*), and stuffed toys and theme park characters (*plushophilia*). The most common fetishes include shoes, leather and latex items, and women's underwear.

How abnormal is this? Is Jordan Haskins just a regular guy with dreams and political ambitions whose curiosity online led him down a big psychological rabbit hole?

Like lots of other atypical behaviors, fetishes exist on a continuum, in gradations from mild—infrequently expressed tendencies—to full-blown regularly manifested behaviors. Many people may recognize or experience mild versions of these desires or "interests" (as DSM-5 delicately puts it), but they emerge only in private fantasies. If an individual has a specific fetish and expresses it with a willing partner, or the behavior is integrated into normal sexual activity, then it is not considered a disorder.

Most fetishes are not harmful to anyone, but there can be legal implications. Interestingly, the crime most associated with the behavior is burglary—committed by an individual who needs to satisfy the urge for a fetish object, like a woman's bra or panties, and breaks into a neighbor's house to steal it. In the same way, what landed Jordan Haskins in jail wasn't cranking; it was trespassing and pulling spark plugs, which constituted damage to government property. The specific misdemeanor charge was "unlawful use of a motor vehicle."

A fetish is considered a disorder if it provokes intense recurring fantasies and if partners are obligated or coerced to participate. Then it is considered behavior that can cause problems in a relationship and may escalate to becoming all-consuming and destructive. In clinical terms, the difference between a sexual preference and a diagnosable disorder comes down to one key word: *distress.*

To be diagnosed with a paraphilia disorder, a person must:

- feel personal distress about their interest, not merely distress resulting from society's disapproval or the disapproval of significant others;

or

- have a sexual desire or behavior that involves another person's psychological distress, injury, or death, or a desire for sexual behaviors involving unwilling persons or persons unable to give legal consent.

I am often asked why politicians and celebrities give in to this kind of behavior with such apparent frequency. My answer is that public figures with careers that thrive by amplifying their presence online are probably more exposed and noticeable. When you are standing on a stage, a much larger crowd will be seeing your act. I believe their struggles simply mirror the struggles of many others. Technology can make it harder for anyone to control impulses and can facilitate and escalate problematic behavior.

The Normalization of a Fetish

Over the centuries—and probably throughout the history of humankind—there's been a small but consistent human interest in sadism as it relates to sex. The word *sadism,* or arousal from inflicting pain on another person, refers to a bisexual French aristocrat, libertine and prolific author, the Marquis de Sade—or Donatien Alphonse François de Sade—who died in 1814 and left behind a slew of erotic literary novels, short stories, plays, dialogues, and political tracts. A proponent of extreme freedom and unrestrained morality, the marquis spent years in prisons and asylums because of his own appetites and behavior, which his works are based on, and in which sexual fantasies involve criminality, blasphemy, and violence. Most of his works were suppressed until the mid-twentieth century, when the Marquis de Sade became a subject of interest to intellectuals, who called him, in turn, a nihilist, a satirist, and a precursor to Freud and existentialism. The French poet Guillaume Apollinaire described de Sade as "the freest spirit that has yet existed."

Pain is the central interest of sadomasochism, either inflicting it or receiving it. And it appears to have ongoing appeal for a small percentage of the population. Over time, there has been a steady underground interest—but not quite acceptable for dinner-table conversation—in sadomasochistic "play" between consenting adults. In 1953 the pioneering American biologist and sexologist Alfred Kinsey determined that 12 percent of women and 22 percent of men admitted to responding sexually to sadomasochistic narratives; twice as many males and almost equal numbers of women were found to have responded erotically to actual pain (in the form of being bitten). The results led Kinsey to the conclusion that "males may be aroused by both physical and psychologic stimuli, while a larger number of the females, although not all of them, may be aroused only by physical stimuli." A study done two decades later, in 1974, in the midst of the so-called sexual revolution, found that 5 percent of men and 2 percent of women reported that they obtained sexual gratification from inflicting pain.

An early study of Canadian men in 1976 found that 10 percent had sadomasochistic sexual fantasies during sexual intercourse, and a more recent study of Canadian women in 2008 found that a large percentage—from 31 to 57 percent—were reported to have rape fantasies. These are reported "fantasies," not reported experiences. For 9 to 17 percent of women, these were a frequent or favorite fantasy experience. Another study found that 33 percent of women and 50 percent of men had sexual fantasies of tying up their partner, although the participants were not asked if they'd ever done this. According to these studies, a fairly significant number of people are aroused by thinking of these scenarios or have participated in them to some degree, either once or repeatedly.

What else does this information tell us? It provides a possible scientific explanation for the wild popularity of the novel *Fifty Shades of Grey*. Since its release in 2011, and the publication of its many sequels, it has become the bestselling book of all time excluding the Bible, with sales now exceeding 100 million copies worldwide. That is a big number. And a lot of readers.

I guess most people on the planet know by now that the book tells the story of a young entrepreneur, Christian Grey, and a literature student, Anastasia Steele, who become romantically involved, but only on Grey's terms—which means participating in bondage, dominance, and sadomasochism, which is referred to nowadays as BDSM. The book introduced a couple of serious paraphilias, *sexual sadism disorder* and *sexual masochism disorder,* to the general reading public as a fun and fascinating pastime.

Certainly there have been other attempts to mass-market paraphilia—from Madonna's bestselling coffee-table book, *Sex,* in 1992 to *Belle de Jour,* the 1967 Luis Buñuel classic film about a young woman who engages in fetishistic behavior. Catherine Deneuve stars as the newlywed who's obsessed with her father and fantasizes about whips, crops, domination scenarios, and bondage. Even though she really loves her new husband, a nice doctor, she can't bring herself to be intimate with him. To seek pleasure and fulfill her fantasies, she begins to spend midweek afternoons as a prostitute while her husband is at work. And who could forget the BDSM scenes in Quentin Tarantino's 1994 film *Pulp Fiction*—or the line "Bring out The Gimp"?

It's fascinating to read about rebels and libertines. They deliver vicarious thrills and fulfill our fantasies of living large and wild—and not caring what society thinks. The "forbidden" and even disturbing aspect of *Fifty Shades* is clearly part of its popularity. Curious to see what this book offered the general public, I launched into my first *Fifty Shades* novel with some trepidation. In full immersive ethnographic research mode, I set out to read the trilogy in order to critique it, but after sampling thirty or forty pages of the first installment, I stopped hunting for irresponsibly lighthearted treatment of a serious mental disorder and found myself worrying about the psychological disposition of the writer, E.L. James. But then, I always do that.

Another ten pages and I had to give up. It's hard enough to study reports of actual clinical psychopaths or the criminally insane. Reading fictional accounts of torture and brutality was, for me, even more disturbing. But my greatest concern is the underlying message: This behavior might be fun, or even romantic.

Twenty or thirty years ago, a person with a fetish or guilty pleasure of his or her own had to dig around in the public library for a copy of the Marquis de Sade's writings, go to an art-house cinema to see *Belle de Jour,* or go to a theater that featured pornographic films. Those in dire need of this material could turn to commercially available pornography. How difficult was it to find? An analysis of the covers of heterosexual porn magazines in the 1980s showed that more than 17 percent depicted bondage and domination imagery. In other words, even in porn magazines, there was limited access to this type of enticing material. The potential for escalating and reinforcing the behavior was somewhat limited. Now technology has changed that.

Is it a coincidence that in the years following the publication of the Fifty Shades trilogy, Internet searches for BDSM porn worldwide have risen by 67 percent and searches for terms like "sex slave" and "master" have increased nearly 79 percent and 72 percent, respectively? Membership in FetLife, a Vancouver-based pansexual social-networking site that serves the BDSM lifestyle, has more than tripled; there are more than 3.5 million members of the FetLife community, and they've shared more than 19 million photos and 172,000 videos, participated in 4.7 million discussions, and created 1.7 million blog posts.

FetLife describes itself as "similar to Facebook and MySpace but run by kinksters like you and me. We think it's more fun that way. Don't you?"

There are risks in practicing sadism for sexual pleasure—and obviously part of the excitement is the risk itself. But the main thrill comes from inflicting pain, apparently using an array of implements: paddles, wooden spoons, electric cattle prods, skewers, and knives, along with the traditional whips. According to an article by William Saletan in the online magazine *Slate,* even women who have appeared in BDSM pornography have reported being injured in the production of the scenes—receiving electrical burns, injuries requiring surgery, and permanent scars from beatings.

Saletan goes on to say, "While these injuries were accidental, the BDSM subculture doesn't regard intentional harm as wrong. According to the 'Statement on Consent' developed by the National Coalition for Sexual Freedom, injury is wrong only if it 'was not anticipated and consented to.' "

To me, this seems an unusual notion of freedom: *the right to be hurt.*

Finding Cohorts Online

Sadomasochistic relationships run the range from mild to severe, depending on the power/control needs and expression of passivity/vulnerability. There is *comorbidity* between sadism and psychopathy, which means that these two traits are often found together. But let's remember there are also a significant number of studies in the past thirty years showing that individuals who practice sadomasochism demonstrate evidence of good psychological and social function, as measured by higher educational level, income, and occupational status compared with the general population. In one of these studies, U.S. sociologist Dr. Thomas S. Weinberg concluded that "sociological and social psychological studies see SM practitioners as emotionally and psychologically well balanced, generally comfortable with their sexual orientation, and socially well adjusted."

While an interest in BDSM doesn't necessarily point to larger problems, given the explosion of BDSM lifestyle material online in the past ten years, and the normalizing of the behavior generally, I have to wonder about the overall impact this has on vulnerable people—those with mental health conditions and disorders. There have been troubling stories and tragedies along these lines, related to how quickly partners for BDSM scenarios can now be found. Masochists sometimes say that it's hard to find a "good sadist"—or one who understands their preferences. The Internet comes in handy in terms of BDSM matchmaking, just a few keyboard clicks away.

Prior to the invention of the Internet, finding a willing partner or a group to participate in a BDSM scenario wasn't so easy. I have been working on a theoretical construct I call *online syndication,* which is really about the mathematics of behavior in an age of technology. It works like this: If I'm a sex offender in a small town in North Dakota and you are a sex offender in a small town in rural Georgia, what are the chances of us meeting each other in the real world?

This used to be capped or bound by the laws of probability and domain. In other words, it was restricted by chance and proximity. Two sex offenders who lived so far away from each other had very little if any chance of ever meeting. Now that has changed—not just for sex offenders, but for girls with eating disorders, cybercriminals, and people with fetishes. All of these groups can easily syndicate to socialize, normalize, and facilitate their particular interests. I hope I'm wrong, but I believe this cyber effect could result in a surge in deviant, criminal, and abnormal behavior in the general population.

You can join communities like FetLife, a dating site like Alt.com, or domsubfriends, a BDSM education and support group. There is also Tabulifestyle (also known as TLS), a matchmaking service for people with taboo tastes. Describing itself as "a service that caters to sexually adventurous couples and singles," TLS claims to offer "a secure and discreet community for 'REAL' members who are sexually open-minded. Tabulifestyle's comprehensive set of profile features, unlimited picture galleries, robust search functionality, members and swinger clubs calendar, advanced blocking and filtering tools provides a FUN safe and secure environment."

How FUN? How safe?

You don't have to look far for sad and disturbing examples of people whose paraphilia escalated and amplified after they found cohorts on a social-networking site. In 2012, a child-care worker in Ireland, Elaine O'Hara, disappeared, and the media began to closely follow the case. Several unusual things were found at the single thirty-six-year-old woman's apartment, including a latex bodysuit and images of two hunting knives. O'Hara, who suffered from suicidal thoughts and depression, had a history of cutting herself. Her father's partner, a psychologist, had assessed Elaine's emotional age as equivalent to a fifteen-year-old's. Elaine had told her father that she had met someone who enjoyed tying her up and that she had asked him to kill her.

A year later, in 2013, her remains were found in the underbrush in the Dublin Mountains by a dog walker. Not far away, submerged in the muddy bottom of a reservoir, her mobile phone was found—along with a rusted chain, a bondage mask with zips over the holes for the eyes and mouth, a length of rope, knives, and other BDSM paraphernalia. Retrieval of her text history revealed O'Hara had a relationship with a man who declared himself a sadist.

In exchanges, he and O'Hara expressed an interest in stabbing and being stabbed for sexual gratification. One message to her read: "My urge to rape, stab or kill is huge. You have to help me control or satisfy it."

Graham Dwyer, a forty-two-year-old Dublin architect and father of three, was arrested and charged with O'Hara's murder. In 2015 he was found unanimously guilty by a jury and jailed for life.

So how did the murderer and his victim find each other? O'Hara left behind a notebook containing the name of a lifestyle community she was involved in: FetLife. Dwyer was reported to have used FetLife, Alt .com, and other BDSM websites to indulge his fantasies. Over the course of the couple's online/offline relationship, thousands of profoundly disturbing text messages passed between them, which were reportedly backed up on her laptop. In the master-slave language they used with each other, O'Hara revealed the mental distress she suffered throughout their BDSM affair.

He was disinhibited enough to indicate his profession in his online alias, "Architect77." Their text messages show escalation, from discussion of BDSM to talk of murder. O'Hara had other partners she had met online, and their perceived anonymity was shattered when they were summoned to give evidence in the trial. But mostly the story of Dwyer and O'Hara is about online syndication—outliers clicking to connect. A master was looking for a slave. A slave for a master. A woman with a history of cutting herself met a man who had a fetish for stabbing.

O'Hara was a fan of TV crime dramas, like *CSI,* and ironically she had warned Dwyer of the dangers of being caught for murder through DNA and cellphone logs. "Technology is a killer now Sir," she texted him, more than a year before she was killed.

Cyber-Socialization

As bio-psychologist Bruce King has written, when it comes to sexuality, what's "normal" depends on where you are. In cyberspace, we know that people may do and say things that they wouldn't do in the real world, due to the effects of anonymity and online disinhibition. This environment plays a significant role in socialization as well. All geographical barriers are removed when we connect with others online. And with limited social cues, as I've discussed, in cyberspace we make friends and meet new people without the help of our real-world instincts.

This has both positive and negative results. The magnificent upside of the barrier-free connection is that we can make friends with people we'd otherwise never know—gain insights into other lives and situations. This is socially broadening and educational, and also generates empathy and understanding. For individuals who are socially isolated—due to distance or personality issues—there's now a place to reach out and find meaningful relationships. For young people anywhere who might feel lonely and curious about their sexuality, the Internet offers a way to explore. Adolescence is an age when experimentation occurs anyway, so some would argue that curiosity and experimentation online could be better and safer than in the real world. (I'll be discussing this in greater detail in chapters about teens online, and again in cyber-relationships.)

The downside: We can blindly fall into dubious friendships and social connections. Online syndication isn't just about finding other people who share your interests. It can ignite a process of norming and socialization that, I believe, when it comes to deviant or criminal behavior, presents an enormous threat to society if not recognized or mitigated.

The *Oxford Dictionary of Psychology* defines *socialization* as "the process, beginning in infancy, where one acquires the attitudes, values, beliefs, habits, behaviour patterns, and accumulated knowledge of one's society . . . and modification of one's behavior to conform with the demands of the society or group to which one belongs."

Here's how it works: A group or community assimilates new members by familiarizing and educating them in its ways. Online, familiarization can be formal or informal. Norms and rules can be communicated explicitly or implicitly. Successful socialization is marked by acceptance. In social psychology we call this "norming." If you have been involved in a group of any kind, you probably noticed that as members start to bond, a group identity forms. This is part of the norming stage of group development, which is a natural part of socialization.

What changes online?

Cyber-socialization happens much more quickly because we are hyper-connected. Online communities and networks are built on the foundation of individuals, or *actors,* as we call them in cyberpsychology, online contacts who are friends, close friends, collaborators, or colleagues and connected by ties—that is, relationships or special interests. Gardeners find each other on gardening forums. Cooks find each other on food sites. Their "tie" is their special interest in gardening or food. But that tie can become very specific almost immediately—and you find yourself in a community that's interested in cooking with parsnips or the spice fenugreek. The more specific the tie, the stronger the bond. In terms of atypical sexual preference online, the fetish is the tie.

While the popularity of a book like *Fifty Shades of Grey* normalizes bondage, dominance, and sadism—so it is no longer a taboo or forbidden subject—an online Web community devoted to this practice will socialize these fetishes. In other words, the popularity of the book makes it "okay" for you to show an interest in BDSM and feel comfortable browsing the various websites devoted to the subject. Once you begin interacting with members or join a community, you are brought into a belief system. This means that you might adopt the attitudes, values, beliefs, habits, behavioral patterns, and accumulated knowledge of the society you've joined. In social psychology this is called one's "reference group."

This norming process can encourage further explorations and adventures too, which are more likely to happen given the powerful force of anonymity online. Sharing your stories in a community like this can be competitive and may lead you to behavior that is even riskier, almost as if you were dared.

But let's apply curiosity and experimentation to paraphilia, some of which can become quite compulsive. Say you're curious and searching online—and come across communities and practices that are new and interesting. Over time, as you are cyber-socialized in this community, you can adopt the belief system of the group. What may be initially troubling, or make you uncomfortable, can seem normal over time.

The common theme in the BDSM scenario is pain and discomfort. A person with a masochistic disorder is sexually aroused by the act—real or simulated—of being beaten, humiliated, abused, or tortured. Sometimes this is just verbal humiliation, but for some it means self-inflicted cuts, burns, and piercings. Masochistic sexual activity can involve simulated punishments, like spanking, or rape. The problem with these behaviors is that they can become escalated—more and more extreme versions and scenarios are required to cause the desired result. There are aspects of BDSM that are compulsive, even addictive and destructive, and some people may be more susceptible to the cultlike trap of escalating behavior. I can't help but wonder if this could happen more quickly online, due to the combined cyber effects of socialization, syndication, escalation, and online disinhibition.

And what about individuals who are suffering from sadistic behavior disorder? Does the mainstream popularization of BDSM encourage and normalize more extreme behavior? Does it mean a greater pool of willing experimental partners? It doesn't surprise me that the U.K., which has taken a progressive role in online governance, has amended its regulations for paid-for video-on-demand films and now bans images that depict abusive, violent, and sadistic behavior—such as can-ing, aggressive whipping, spanking, and face-sitting, as well as life-threatening acts such as strangulation. I believe if there were more such consideration of ethics in cyberspace, greater governance, better education, and, if necessary, appropriate regulation, it could spare many vulnerable individuals from harm and pain and prevent susceptible people from going deeper into behaviors that may ultimately be destructive. Great societies are judged not just by how they serve the strongest but by how they protect the weakest and most vulnerable. We need to collectively focus on creating the best possible cyber society. Pursuit of the greater good should never go out of fashion.

Cyber-Exhibitionism

The Internet is like a catalog of desire begging people to flip through it. Think of the laboratory experiment in which the men watched an erotic slideshow and wound up with a fetish for boots. Now think of the erotic slideshow that is the Internet, and what sorts of new desires, and new behaviors, are being created.

It's hard to forget Anthony Weiner, the skinny, superambitious American politician who posted photos of his genitals to a selection of different women online while engaged (and later married) to Huma Abedin, a talented and attractive woman with a highly visible political job. *What is wrong with this picture?* Weiner, an otherwise accomplished individual—serving the 9th District of New York for thirteen years—was forced to give up his congressional seat following a sexting scandal that consumed the United States for weeks in 2011 (the same year that *Fifty Shades of Grey* was published, interestingly enough).

Exhibitionistic disorder is a mental health condition, a paraphilia that centers on a need to expose one's genitals to others, typically strangers caught off guard, in order to gain sexual satisfaction. Men make up the vast majority of people who participate in exhibitionism. And nearly all targets of exhibitionism are women, underage girls, or underage boys. Usually the behavior begins during the first decade of adulthood, although some individuals do start later in life. Roughly one-third of all men arrested for sexual offenses in the United States are exhibitionists.

Flashers are a subject of lots of jokes and humor, cartoons and comic sketches. But exhibitionism is a real disorder, and it would be compassionate to remember that people may not choose this way of life. Some psychologists believe it is driven by profound feelings of personal inadequacy. The exhibitionist may be afraid to reach out to another person out of fear of rejection and is led to exhibitionism as a way to somehow involve others, however briefly, in an intimate moment. Logically, if contact is limited to just the opening of a raincoat before dashing off, or the quick snapshot of one's private parts sent to the in-box of an unsuspecting woman, the possibility and pain of overt rejection are minimized.

Some men expose themselves looking for affirmation of their masculinity. Others may simply seek attention they crave. Anger and hostility toward people, particularly women, may drive some exhibitionists. In this case, they expose themselves to cause shock and frighten.

Like many paraphilias, exhibitionism is difficult behavior to give up because it's typically a source of great excitement and pleasure. Most people are motivated to continue, which is why treatment for exhibitionism, as well as many other paraphilias, is a complex process, and several methods are usually tried without success. Exhibitionists have the highest rate of rearrests of any sexual offender.

As far as we know, Weiner was not an open-raincoat type—a guy who flashed women on the street in order to shock them. Rather, he befriended female strangers online and quickly transformed informal chitchat about politics and policy into a sexually explicit exchange—unprovoked and unwanted sexts.

The practice of sexting, or exchanging intimate text and images online, is increasingly popular behavior, almost normalized, but nonetheless it's an awfully high-risk practice for a public figure like Weiner. In his case, the behavior was so reckless, it defies logic. I can't help but wonder what other factors were at play—and what he hoped to gain that was so important to him in the moment. Was it a need for power or to shock women? Or both?

A recent study at Ohio State University demonstrated that men who post a lot of selfies—particularly if they were edited or Photoshopped beforehand—scored higher on measures of narcissism and psychopathy than men who didn't. Narcissism is the belief that you're smarter, more attractive, and better than others. Psychopathy is characterized by egocentric and antisocial activity. Follow-up work suggests that the same findings apply to women. Both the narcissist and the exhibitionist are hungry for feedback or some reaction to their behavior. There is a self-reinforcing cycle: When a selfie is posted, it leads to feedback, which encourages the posting of more selfies. "We are all concerned with our self-presentation online," said Dr. Jesse Fox, the lead author of the Ohio State study, "but how we do that may reveal something about our personality."

Weiner revealed a lot more than personality. There are still a few of his explicit selfies available online—the staying power of images online is another reason why this kind of behavior is so risky for public figures—and an interesting study could be made of them, in terms of what's called "content analysis," or the forensic analysis of the content of an image, typically the subject, pose, environment, identifying objects, and other details that can provide a lot of information.

I have conducted extensive research in image-content analysis, and I'm now in the process of developing a software tool that will help police to extrapolate more data and information from an image. For this tool, I have organized content analysis into five categories, ranging from demographic detail (age, gender, and ethnicity) to situational (identifying objects, environment, and setting). I have developed a grid system to apply to each image that breaks it into segments to be analyzed individually. Each segment can be zoomed into and methodically investigated, much like a systematic room-to-room search done by police.

I used my systematic approach to analyze one of Weiner's sexts that was cropped to make it fit for publication in the New York *Daily News,* and it's fascinating how many identifying details there are. In a classic "power shot," he shows off his bare torso and flexed pecs, and behind him, on a narrow table over his right shoulder, is an array of framed personal and professional photographs. Was this unintended? A person taking a selfie can be either disinhibited—lost in the moment and oblivious of surroundings—or quite conscious and therefore carefully staging the image, and designing it to impress. In forensics, "staging" means deliberately falsifying a crime scene (when the offender alters the evidence), but in this case, I am using the common definition of *staging*—setting a scene in order to make a certain impression.

Once again, the key word used to distinguish between a behavior and a true disorder is *distress*. Clearly Weiner's behavior was persistent enough, and distressful enough, to destroy a political career that he had been diligently polishing and honing for decades. What would drive a man to such risky behavior and cause him to become an object of such incredible ridicule? Unfortunately for him, his name still elicits laughter.

I call this *cyber-exhibitionism*. It appears to be a mutation of real-world behavior and part of a new generation of paraphilia in cyberspace, where there is better reach, a wider audience, more victims, compromised judgment, more risk-taking, heightened distress, and, most important, permanent digital records. For an individual with the sorts of apparent needs and vulnerabilities that Weiner demonstrates, the forces of online disinhibition, escalation, and impulsivity are enormously powerful. And enormously destructive. One of the earliest forms of cyber-exhibitionism, "JenniCam," came out of a university art project. A young woman broadcast from her dorm room 24/7, which eventually escalated to her allowing viewers to watch her nude, then to allowing viewers to watch her having sex.

The nature of the relationship between voyeur and exhibitionist is symbiotic, almost parasitic, and explains the phenomenon of reality TV. In a later chapter on teenagers and technology, I will discuss sexting or nude selfies as part of the courtship ritual among young people, and also a mild form of cyber-exhibitionism.

A bizarre case in the U.K. recently showed how criminal and deviant behavior, facilitated by technology, is continually evolving. In 2015, Lorraine Crighton-Smith was traveling on a train in South London when her iPhone was suddenly bombarded with explicit photos of a man's genitals. The thirty-four-year-old woman, shocked by the first image, instantly declined it. As soon as she did, another image appeared. Then she realized that she had switched on "Airdrop" on her phone to share photos with a friend who was also an iPhone user. By mistake, her privacy setting for Airdrop was put on "everyone" instead of "contacts only." This allowed her phone to be accessed by someone sitting nearby on the train, which takes the new term *cyber-exhibitionism* to a new place.

Two years after his sexting scandal, Weiner announced his candidacy for New York City mayor, hoping that the electorate had moved on—or forgotten. But the Internet is unforgiving. More sexts surfaced. To stifle more controversy, he quickly acknowledged that he had continued to cyber-flash and send explicit images online to at least three women in 2012. Hoping to stay in the race, he and his wife appeared at a press conference together, where he apologized: "I want to again say that I am very sorry to anyone who was on the receiving end of these messages and the disruption this has caused."

Refusing to drop out, he hung on—and forty-eight hours before the primary, he mused, "Maybe if the Internet didn't exist . . . if I was running in 1955 . . . I'd probably get elected mayor." Instead, he lost the mayoral primary with less than 5 percent of the vote.

On his disastrous election day, when reporters asked Weiner what he planned to do next, he whipped out his middle finger and flipped the bird. Perhaps the next best thing to the open raincoat.

Webcams + Cyber-Voyeurism

Another paraphilia that has migrated and morphed significantly online is *voyeurism,* also known as *scopophilia*—or the recurrent preoccupation with fantasies and acts that involve observing persons who are naked or engaged in grooming or sexual activity. What once was simply the classic "peeping Tom" has been impacted by technology in the past century by the invention of the camera.

A case of this in the real world involved a prominent rabbi in Washington, D.C., who hid a small camera inside the ritual bath of the National Capital Mikvah, next to the Kesher Israel Congregation in Georgetown. When female congregants came to the basement bath and shed their clothes to practice the ancient sacred purification rite of dunking in the water and reciting a blessing, a small digital camera hidden inside a clock radio took their pictures.

In 2014, Rabbi Barry Freundel, renowned as an authority on Jewish law and ethics and an "intellectual giant," according to the *Washington Post,* was charged with six counts of voyeurism and faced up to six years in prison. Eventually, he pleaded guilty to fifty-two counts of voyeurism. Prosecutors identified a further one hundred recordings of women, but these fell outside the statute of limitations. In May 2015, Freundel was sentenced to six and a half years in prison.

Over the past few decades, there have been cases of this kind of privacy violation—cameras hidden behind hotel-room mirrors, cameras hidden in women's bathrooms. The lawmakers hoping to protect individuals from privacy violations have always raced to keep up with technological changes. Imagine if a voyeur with a persistent disorder like Rabbi Freundel's found ways to access the webcams of a young woman's computer? What if he were able to access dozens of these webcams at once?

This brings me to our last story of the chapter.

Five months before Cassidy Wolf, of Temecula, California, was crowned Miss Teen USA in 2013, the poised and picture-perfect nineteen-year-old received an anonymous email from a man who had hacked into her computer and gained control of its webcam. With it, he had spied on Wolf in her bedroom. How?

The voyeur had surreptitiously installed malicious software on Wolf's computer, using some form of Remote Access Trojan, or RAT, that can be bought online for as little as forty dollars. He told Wolf that he would release images that he'd gotten from her webcam unless she complied with one of his three demands: send nude photos, send a nude video, or log on to Skype and do whatever the hacker asked for five minutes.

No dummy, Wolf contacted authorities immediately, and an FBI investigation was launched. Three months later, Jared James Abrahams, a nineteen-year-old man who had gone to high school with Wolf, was arrested. He had gained control of twelve different women's webcams and had hacked into an estimated 100 to 150 other computers, and was sentenced to eighteen months in prison.

Raising awareness of cybercrime became Wolf's mission during her year as Miss Teen USA. "I wasn't aware that somebody was watching me," she told an interviewer. The light on her webcam hadn't even gone on. While traveling the country, she offered tips for cyber-security: change passwords frequently, delete browsing history regularly, and—most important—put a sticker over the computer's webcam lens when you're not using it.

The webcam may be technology's greatest gift to the voyeur. Considering that more than 70 million computers were purchased in the United States in 2015 alone, not to mention the preexisting devices—the smartphones, tablets, and desktops with cameras that connect to the Internet—it's daunting to imagine the spying capabilities of voyeurs.

Previously Unimaginable

A new sexual freedom has been spawned by the Internet—encouraged by the effects of anonymity, cyber-socialization, online syndication, disinhibition, and escalation—and even given rise to new or formerly unknown fetishes like cranking. The cyberpsychological reality: One can easily stumble upon a behavior online and immerse oneself in new worlds and new communities, and become cyber-socialized to accept activities that would have been unacceptable just a decade ago. The previously unimaginable is now at your fingertips—just waiting to be searched.

Technology isn't the problem. It's that we don't yet know the full effects of the cyber environment or where it is taking us. Like the beginning of many kinds of life adventures, sexual exploration on the Internet can be exciting at first. You may tell yourself you are just dipping your toe in the water. But what if that water feels great—*just wonderful.* And what if, soon afterward, all you can think about is getting your toe back in that water again. Before long, you may be bouncing on the diving board and jumping in. How can you resist?

And when is it time to stop?

CHAPTER 2

Designed to Addict

Soon after Alexandra Tobias, a twenty-two-year-old mother in Florida, called 911 to report that her three-month-old son, Dylan, had stopped breathing and needed resuscitation, she told investigators that the baby was pushed off the sofa by the family dog and hit his head on the floor. Later, full of regret and sorrow, she confessed to police that she was playing *FarmVille* on her computer and had lost her temper when little Dylan's crying distracted her from the Facebook game. She picked up her baby and began shaking him violently, and his head had hit her computer. At the hospital, he was pronounced dead from head injuries and a broken leg.

At the time of the 2010 incident, *FarmVille,* a wildly popular online game where players become virtual farmers who raise crops and livestock, had 60 million active users. Described in glowing terms as "highly addictive" by its fans, there was eventually a need for FAA (*FarmVille* Addicts Anonymous) support groups, and even an FAA page on Facebook itself. Can we say that Alexandra Tobias was addicted? Is the explanation that simple? Her virtual cattle were doing fine, but her real life was in ruins.

During her trial, she pleaded guilty and showed great remorse, and in a statement said that she hoped to attend college and make something of herself someday. Her own mother had died recently, she said, and Tobias hadn't felt like herself ever since. She received the maximum sentence in Florida for second-degree murder: fifty years. She'll be in jail for most, if not all, of the rest of her life.

As a forensic cyberpsychologist, I am interested in this sad and disturbing case for one reason: the role of technology in the escalation of an explosive act of violence. In a nutshell, that is extreme *impulsivity,* an unplanned spontaneous act. And in this case, with devastating consequences.

We are all impulsive to a degree. Some people are by nature more spontaneous than others, more likely to act on a whim without too much thought, whether the behavior is driven by joy or anger. One of the beliefs of our culture is that people reach the end of their lives and wish they'd taken more chances and risks. This may be true for restrained, risk-averse individuals. But extremely impulsive individuals would probably say the opposite: It's not the things they didn't do that they regret. It's the ones they did.

There are risks that reward us and risks that ruin us. The same goes for the hours we spend online. This chapter will discuss many aspects of the Internet that are irresistible—whether it's multiplayer gaming, email checking, social-network posting, or bidding on an auction site. Given a host of cyber effects, we may sometimes feel like slaves to our impulses. Why?

The Scale of Impulsiveness

What is impulsivity? It is defined as "a personality trait characterized by the urge to act spontaneously without reflecting on an action and its consequences." The trait of impulsiveness influences several important psychological processes and behaviors, including self-regulation, risk-taking, and decision-making. It has been found to be a significant component of several clinical conditions, including attention deficit/hyperactivity disorder, borderline personality disorder, and the manic phase of bipolar disorder, as well as alcohol and drug abuse and pathological gambling.

Researchers studying attention and self-control often assess impulsiveness using personality questionnaires, notably the common Barratt Impulsiveness Scale, which has been used for the past two decades and was updated in 2014 by a group of researchers at Duke University. It's a really fascinating area of interest—and the test, a list of thirty simple statements to be agreed or disagreed with, can be taken in a matter of ten or fifteen minutes. The statements are easy to answer: "I plan tasks carefully," "I am happy-go-lucky," "I am future oriented," "I like puzzles," "I save regularly," and "I am restless at the theater or lectures." You can find the entire test online, but the results need to be assessed by a professional, so taking it on your own, you won't wind up with a final score. Reading over a few of the statements will give you some insight into the three types of impulsivity, which can be physical (difficulty sitting still), cognitive (difficulty concentrating), and sensory (difficulty resisting sensory rewards).

A highly impulsive individual—in all three aspects of impulsivity—tends to be restless, happy-go-lucky, uninterested in planning ahead or saving, not future-oriented. In common parlance, he or she might be described as ADD or ADHD, but it's a bit more complicated than that. When a child is diagnosed with an *executive function disorder,* like ADHD or other attention-related problem, one of the likely aspects of this is something described as suppressed response inhibition, which is generally defined as the inability to suppress an urge to do something, even when environmental contingencies demand it. In other words, the world is telling these children not to do something—"Don't stand up on the bus!"—and telling them why it's a bad idea, but their ability to restrain themselves just isn't there.

A person who is diagnosed as having obsessive-compulsive disorder shares the trait of impulsivity and suppressed response inhibition with ADHD, but he or she finds urges extremely hard to control—and stop. This same trait has been observed in alcoholics, cocaine addicts, heroin- and other substance-dependent patients, as well as smokers. Recent studies have also found impulsivity to be positively correlated with excessive computer game playing and excessive Internet use in general.

Before I go into this subject any deeper, I want to discuss the difference between *impulsive* and *compulsive.* In everyday conversation, we tend to use these two terms almost interchangeably, as if they meant the same thing. But they are actually at opposite ends of a spectrum of behavior. While impulsive behavior is a rash, unplanned act, such as Alexandra Tobias's rage at being interrupted while playing *FarmVille,* compulsive behavior is planned repetitive behavior, like obsessive hand washing or cranking, as discussed in the previous chapter about paraphilia.

Let's describe these in cyber terms. When you constantly pick up your mobile phone to check your Twitter feed, that's compulsive. When you read a nasty tweet and can't restrain yourself from responding with an equally nasty retort (or an even nastier one), that's impulsive.

What makes the Internet so alluring? Why do some individuals struggle more than others to pull themselves away from their mobile phones and computer screens?

Fun Failures

Why does anybody buy a Powerball lottery ticket if they know the chances of winning are one in 300 million? For some of the very same reasons that keep people playing any number of beguiling online activities. When we invest time playing *League of Legends* or spend money on the lottery, we know that there's little chance of "hitting it big." Sporadically, though, small rewards do come, and these *intermittent rewards,* as they're called, bring us back again and again.

It's an accepted fact in behavioral psychology that intermittent reinforcement is much more effective at motivating people than continuous rewards. If you are rewarded randomly for an activity, you are likely to continue doing it—far more likely than if you are rewarded each and every time. A famous study of pigeons demonstrated this: When pigeons were consistently rewarded for a certain activity, they did not necessarily continue the activity. But they were much more responsive, and much more prone to act, when given intermittent reinforcement. Maximum responsiveness was achieved when they were rewarded half the time.

Here's how it works with scratch-off tickets: You are asked, say, to scratch off six squares on a card and reveal the numbers hidden underneath. You will win if you uncover three matching numbers. In terms of rewards, just the act of scratching off the hidden numbers is a little exciting. A drama is unfolding, and that creates expectancy, which has been found to deliver a little dopamine to the brain. Dopamine is an organic chemical released in the brain that helps us regulate movement and emotional responses, and is also associated with pleasurable feelings. More than 110,000 research papers have been written about dopamine in the past sixty years. In pursuit of the pleasure it gives us, we do things that release it. What is fascinating and underresearched is the role of technology in this process.

Most scratch-off cards are designed to lose. But they are also designed with many matching symbols. Why? You scratch the card, a sequence of matching symbols begins to appear, and you get excited thinking that you may win. For a thrilling second, or two, or three, you believe your card is a winner. In the gambling trade, this tease is called "a heart-stopper," because it can give you a surge of excitement, a little buzz of pleasure. This is classic positive reinforcement. So even when you don't win, that temporary buzz of excitement is enough to bring you some pleasure and reinforce card-scratching behavior. And later, the biochemical and psychological memory of that pleasure is enough to keep you in the feedback loop, and buying more lottery tickets. It's not a huge blast of pleasure, mind you, but it's just enough.

This is what conditioning is all about.

When any behavior is rewarded with pleasure, you are more likely to repeat it. The psychology of a casino slot machine works the same way. Three wheels of the slot machine are spinning—and showing you all those matching pairs. Two wheels stop. And if the symbols match, it's a heart-stopper moment. The third wheel stops, and you lose. But somehow, it felt fun anyway.

In game design this is called "fun failure." Even though you are failing miserably, you aren't miserable. Why? That biochemical pleasure hit makes all the difference. And the mere act of anticipating winning is fun. This is what keeps people buying lottery cards, feeding a slot machine, or playing *Candy Crush Saga*.

Who hasn't felt the draw of the cyber fun-failure vortex? Who hasn't wasted time or money or both online and still managed to feel it was fun? There is more to it than heart-stoppers. Each type of online activity has its own attractions, extra built-in rewards that condition users to return.

Why is the mere act of searching online so hypnotically compelling? Why are the alerts and notifications on a mobile phone impossible to ignore? Since my interest is in forensic cyberpsychology—and therefore I am a bit more focused on pathological behavior than the average person—I have to look at how the various rewards of being online may have a dark side for some people, and what the implications are for the rest of us.

I Seek, Therefore I Am

If you start simply with psychologist Abraham Maslow's famous "hierarchy of needs"—the needs that demand our attention and motivate human beings to survive, adapt, and evolve—you'll see them all met online in one manner or another: from physiological needs to needs for safety, love, belonging, esteem, self-knowledge, and self-actualization.

Online anonymity offers you a sense of safety. Joining an online community, or participating in a multiplayer online game, can give you a sense of belonging. Getting your Instagram photos or Facebook posts "liked" meets a need for esteem. But that's just the beginning of social-networking rewards and pleasures. According to psychiatrist and author Dr. Eva Ritvo in her article "Facebook and Your Brain," social networking "stimulates the release of loads of dopamine as well as offering an effective cure to loneliness. Novelty also triggers these 'feel good' chemicals." Apart from getting high on likes, posting information about yourself can also deliver pleasure. About 40 percent of daily speech is normally taken up with self-disclosure—telling others how we feel or what we think about something—but when we go online the amount of self-disclosure doubles to 80 percent. According to Harvard neuroscientist Diana Tamir, this produces a brain response similar to the release of dopamine.

Searching online—whether you are hunting down a piece of information, shopping for a pair of shoes, or looking for an old classmate or professional contact—rewards you in another powerful way. Which brings me to a favorite subject of mine, the fascinating real-world work of Washington State University neuroscientist Jaak Panksepp, who coined the term *affective neuroscience,* or the biology of arousing feelings or emotions.

Panksepp conducted laboratory experiments on rats and discovered what he calls the "seeking" system, something that drives both humans and animals to seek information that will help them survive. Dopamine-energized, this mesolimbic seeking system encourages foraging, exploration, investigation, curiosity, craving, and expectancy. In other words, dopamine fires each time the rat (or human) explores its environment. Panksepp, who has spent decades mapping the emotional systems of the brain, calls seeking "the granddaddy of the systems." Emily Yoffe in *Slate* explains: "It is the mammalian motivational engine that each day gets us out of the bed, or den, or hole to venture forth into the world." Seeking is so stimulating, according to scientist Temple Grandin, that animals in captivity would prefer to hunt or seek out their food rather than have it delivered to them.

If we think about this in Darwinian terms, Panksepp is essentially arguing that a number of instincts such as seeking, play, anger, lust, panic, grief, and fear are embedded in ancient regions of the human brain or are, as he describes them, evolutionary memories "built into the nervous system at a fundamental level." To Panksepp, these instincts may be considered adaptive traits so fundamental, and so essential to our survival, that they may even constitute what we think of as our "core-self."

I seek, therefore I am?

You don't need to sing the joys of seeking and exploring to police detectives, investigative journalists, and research scientists. Way before the advent of the Internet, they were experiencing the thrills and rewards of discovery. The drive to seek and explore has kept the human race alive and fed for centuries. But it's Panksepp's work that provides us with a biochemical explanation: The dopamine rewards of seeking and foraging have probably made human beings highly adaptable to new environments. We are rewarded for exploring. One could easily argue that the same reward system, or reinforcement, has made human beings more adaptable to the new environment we are still discovering online.

Addiction is explained in Panksepp's work as an excessive form of seeking. Whether the addict is seeking a hit from cocaine, alcohol, or a Google search, "dopamine is firing, keeping the human being in a constant state of alert expectation." If you think about it, cyberspace is like outer space—infinity in terms of seeking. With our evolutionary memory driving us toward exploring and making sense of this new environment, cyberspace, are we trying to evolve at the speed of technology? And if the biochemical rewards of seeking online are the very ones that can make losing the lottery feel like "fun" to a vast number of the population, what does this mean for individuals who struggle with compulsive gambling or other addictions or ADHD?

Hard to resist. That's how many of us find the Internet. It's always delivering a wild surprise, pulsing with breaking news, statistics, personal messages, and entertainment. The overwhelming evidence points to this: A combination of the fast delivery, exploring opportunities, unexpected information, and intermittent rewards creates a medium that is enticing, exciting, and for some individuals totally irresistible. Now let's add in the design aspects of the apps, ads, games, and social-networking sites—the alerts, push notifications, lights, and other visual triggers that signal us like primitive mating calls.

Check Your Email, Check Your Email, Check Your Email Again. *Now . . .*

The Latin word *addictus* was once used to describe the stretch of time an indentured slave had to serve his or her master. The servant with the sentence was called "the addict."

We've all observed it firsthand: the otherwise polite and well-meaning friend who chronically checks her phone while you're trying to have a lunch conversation. Does she really mean to be so rude?

Her connection with you—and the real world—is competing with the little buzz of pleasure she gets every time she checks her in-box. Most of the emails, texts, or notifications your friend is receiving are not urgent. (Most of her emails are probably advertisements from online retailers!) But she can't stop checking in hopes of getting a personal email from someone she cares about—or exciting news of any kind.

A 2015 study found that Americans check their phones a total of 8 billion times a day. As mentioned in the prologue, a study shows that an average adult with a mobile phone connected to the Internet checked his or her phone more than two hundred times a day. That's about every five minutes. In the evening it escalates. When most people are home from work, on average they begin checking their mobile phones once every six minutes. (How many times have you caught yourself picking up your phone mindlessly and checking your email queue again, then realized you just looked at it two minutes before?) Studies differ, but the overall results are similar: Average phone checking per day is surprisingly high.

Given the dependence on mobile phones, it was only a matter of time before they were repackaged as wearables and strapped to the wrist—but wouldn't this only escalate the distractibility?

In terms of controlling compulsive behavior, there are a slew of tests to take online to scale your own "smartphone addiction." And while these questionnaires are not really scientific, it's worth paying attention if you find yourself starting to feel a little uncomfortable or, worse, nodding in agreement:

- Do you find someone to call as soon as you leave the office or land in a plane? (More important, do you sneak your phone out as soon as the plane has landed and turn it on before the pilot says it's okay? How many can resist that urge?)
- Have you ever been teased because you had your cellphone while working out or doing some other activity?
- Are you unable to resist special offers on the latest cellphone models?
- Do you sometimes believe your phone is ringing, but when you answer it or listen longer, you find it wasn't ringing at all (known as "phantom ringing")?

If you recognize yourself in some of the above compulsive behavior, it might help to understand what makes mobile phones so irresistible. To begin with, they are sleek, well-designed little devils that are portable, they are easy to slip into our handbags and pockets, and they travel with us almost anywhere. (I've heard of swimmers getting waterproof cases for them.) And the mobile phone, like the lottery scratch card, offers an intermittent reward. The surprise of hearing or reading news on our devices gives us a buzz of pleasure, which sets in motion a complex set of reinforcing behaviors: You check your phone to (intermittently) get good or surprising news, which is enough to keep you checking.

Now let's add in what a psychologist would call the *related stimuli* of these digital devices, or the flashing lights and other alerts and notifications that come with each new email or text or Facebook "like," depending on how you've customized your settings. Related stimuli are cues or situations that an addict associates with their addiction. One famous addiction study found that related stimuli associated with drinking alcohol or taking drugs could induce craving, explaining how the sight of a liquor bottle can cause a person to feel the urge to drink. This is the result of classic conditioning, like the lab experiment with the men who became aroused by the slide showing a pair of shoes. In the past, antidrug campaigns often used drug paraphernalia in their posters—syringes, needles, spoons, and piles of white powder, all designed to shock the world into total abstinence. But paradoxically the visual stimuli actually drove some addicts to relapse and led to a fundamental redesign of antidrug campaigns.

Just as substance addicts are constantly fighting urges provoked by related stimuli, the alerts and notifications on a mobile phone can cause its user to have an uncontrollable urge to check his or her device. It isn't so different from the spinning of slot machine wheels or the intense cravings that someone with atypical sexual behavior may have for a fetish object. And while the only noticeable downside for your nice but irritating friend at lunch is that she has alienated you—when will you want to have lunch with her again?—in extreme cases an individual with a serious case of "mobile phone addiction" can become socially isolated and even financially ruined. Depending on where they live and their data plans, compulsive phone users can run up monthly charges that they can't begin to afford.

In the behavioral sciences, a phenomenon called *signaling theory* may help us to understand the irresistibility of mobile phones. Signaling theory, which originated with the study of animal behavior, explains why, for example, peahens choose to mate with peacocks with the biggest tails. Evolutionary psychologists have taken these cues for attention and selection, and have applied signaling theory to understanding human interactions. For instance, a number of research studies have shown that we are likely to be even more afraid of snakes and spiders than we are of large predatory animals such as bears, lions, and tigers. From an evolutionary perspective, this could be because snakes and spiders are difficult to spot, don't make sounds or produce other cues, and are therefore more dangerous. It made sense that our ancestors would look carefully for poisonous creatures before sticking their hands into overgrown brush or putting their feet into moccasins (still a good idea today). Over time, this fear became an instinctive human reaction.

There are several types of signal cues that communicate and attract—visual, acoustic, chemical, and tactile. The visual signals are limited and require a line of sight. A predatory female firefly lures in males with her flashing body light, and then preys upon them, just like the blinking and flashing of your mobile phone. Vervet monkeys have a language of distinct calls representing different types of threats, not unlike the ringtones or your early morning alarm. The waggle dance of the honeybee is a tactile cue to secure social bonds. Next time your phone vibrates in your pocket, you'll feel its need to bond with you. The scent of a queen bee motivates and attracts her worker drones—and no doubt manufacturers are developing chemical signals for their mobile phones. Just wait until your device starts emitting those irresistible pheromones.

Bria Dunham, in an excellent paper on the role of signaling theory in marketing, asks the question "Why are black iPhones sold with white earbuds?" She wonders if perhaps Apple has a backlog of white earbuds or perhaps white ones are cheaper or easier to produce. Eventually, Dunham settles on a more compelling explanation: White earbuds serve a signaling function. Those "telltale white earbuds indicate to passersby that the bearer ascribes to certain notions of coolness and style, engages willingly in some degree of conspicuous consumption, has the necessary resource control to afford a portable Apple device . . .

that's a lot of information content for less than an ounce of plastic and wire."

In other words, unconsciously we might want to display our phones to signal to others that we are part of the Apple Tribe—and have the requisite status and coolness levels to be accepted. This is herd behavior, of course, and there's just as much of this present in real life, and perhaps even more online.

So you've got your phone to prove (unconsciously) that you belong, and then . . . you find that you can't stop checking the darn thing. The problem is great enough that there are now apps created to help compulsive email checkers break these patterns of behavior—or retrain themselves to start feeling "rewarded" by resisting the temptation to check their email in-box. One such technology is BreakFree, an app that will monitor the number of times you pick up your phone, check your email, and search the Web. It offers nonintrusive notifications and will provide you with an "addiction score" every day, every week, and every month—to track your progress. These incentives and rewards help motivate a change in behavior. It's like going on a diet and standing on the scale every night for encouragement.

BreakFree bills itself as a "first of its kind, revolutionary mobile app, aimed at controlling smartphone addiction and helping you maintain a healthy digital lifestyle."

The question is, are you *breaking free* from your compulsion, from yourself, or from the technology? Where does control lie? Who's in charge of your behavior—you or your new app?

Another app called Checky tracks how often you unlock your iPhone and encourages you to share your stats on Twitter and Facebook. It's a spin-off from the app Calm, which sells itself with information about behavioral studies linking compulsive Internet use to ADHD, OCD, and other serious disorders. In the app description, Calm claims to have been created by "recovering" phone addict Alex Tew "to help individuals relax their minds.

> "Like many folks, I am pretty much addicted to my phone," Tew says. "And now I know exactly how much: most days I check my phone over 100 times. In fact, yesterday I checked my phone 124

times. Today I'm at 76, so far. Having this new awareness makes it easier to control my phone usage. My new goal is to check less than 100 times a day."

In psychology, we call this *mindfulness*—adopting Buddhist terminology to describe the state of mind in which our attention is directed to the here and now, to what is happening in the moment before us, a way of being kind to ourselves and validating our own experience. As a way to stay mindful myself and keep track of my time online, I have set my laptop computer to call out the time, every hour on the hour, so that even as I'm working in cyberspace, where time flies, I am reminded every hour of the temporal real world. It's very helpful for me, but a little unnerving for my colleagues who are at the other end of a Skype call and have to hear a voice suddenly call out, *"Eleven o'clock!"*

Some other practical remedies to combat phone distraction—or even compulsive use—are to uninstall some of the beckoning apps on your phone screen. You can also go to your phone settings and turn off your notifications, which are how social media sites like Instagram, Twitter, and WhatsApp keep users checking constantly (because they want you to be checking constantly). Keeping your phone in "airplane mode" will silence it—and prevent you from accessing the Internet. Or you could just go cold turkey every so often and simply turn it off. I went to Bora Bora once and for the first time was in a country where I could not get cell coverage. For the first twenty-four hours I went through the predictable phases of mobile connectivity bereavement: disbelief, anger, panic, and night sweats—followed by exhaustion, then finally *acceptance*. I enjoyed a great five-day break after that, beautifully cellphone free.

For those who are looking for philosophical or intellectual inspiration, a number of books deal with this new aspect of our lives and offer help and insights, including Nicholas Carr's *The Shallows,* Sherry Turkle's *Alone Together,* and William Powers's *Hamlet's BlackBerry.* Since 2006, Powers and his family have taken an "Internet Sabbath," a day or two a week totally unplugged, which he believes has helped them remain mindful, less distracted, and in control of their use of technology. Addiction expert and pioneering psychologist Dr. Kimberly

Young also recommends taking a forty-eight-hour "digital detox" every weekend. Plug your device into its charger and leave it there on Saturdays and Sundays. Even Pope Francis calls for an unplugged Christmas.

The conundrum of "connectivity" is only bound to escalate. More mobile phones are sold each year than the year before. For 2017, the number of cellphone users is forecast to reach 4.77 billion. As the "usefulness" of these devices grows, more people will own them—and will be spending more time on them. We use them to read the news, connect with friends, photograph our lives, shop, manage our address books and calendars, and pay our bills. Meanwhile, we aren't just learning how to use new devices, new apps, and new interfaces. We are learning how to live in a totally new environment—cyberspace—unlike any other we've been in before. When people talk about cellphone addiction, what they could be trying to express is something more serious than just compulsive checking of texts or emails. People feel addicted to technology itself.

What Is Internet Addiction?

It is so memorably ironic that one of the great pioneers of computer and online gaming, Dani Berry, remarked: "No one ever said on their deathbed, 'Gee, I wish I had spent more time alone with my computer.'"

Most studies of Internet addictive behavior—and there have been literally hundreds now—build upon the work of Dr. Young, who has been studying compulsive online behavior since 1994 and had the prescience to open the first Internet addiction clinic in the United States the following year. Young's groundbreaking study compared the addiction-like behavior online with compulsion disorders and found many similarities. Her TED Talk on this subject in 2015 offers more interesting insights and warnings about what she calls the dangers of being "too connected." In research papers and psychological journals, this behavior is sometimes called *Internet use disorder* and *Internet addiction.* As neither of these are formal conditions, I will use the broader term, *Internet addictive behavior.*

In everyday language, the word *addiction* is applied to almost anything that a human being can have a craving for—from eating ice cream to singing in the shower. But to meet the clinical criteria for addiction, there must be a biochemical or chemical component. And for an individual to be diagnosed as having an addiction, they have to experience "withdrawal" and demonstrate a developing "tolerance." In other words, there has to be evidence that an individual has an escalating need—wanting to use the Internet more and more. (That's tolerance.) And when the Internet is removed, it causes distress (withdrawal).

A telephone survey conducted by researchers at Stanford University a decade ago showed a rate of 12.5 percent of the U.S. adult population sample reporting they had "at least one problem" due to overuse of the Internet—often email checking, gaming, visiting cybersex sites, or shopping. The cravings they described were similar to drug and alcohol cravings among addicts. As the years have passed, that statistic—12 percent of the population—seems to have remained fairly consistent, but numbers vary depending on who's doing the research, how the questions are asked, and how "addiction" or "misuse" or "excessive use" is defined. And what is considered "normal use" of the Internet can change from country to country. In South Korea, where the issue of Internet addictive behavior has mushroomed into a much-discussed, much-researched, much-diagnosed, and much-treated condition, studies indicate that about 10 percent of Korean teenagers are Internet addicts. In fact, some demonstrate difficulty in living their everyday lives due to the level of their addiction. Slightly higher numbers have been reported in China, with 13.7 percent of Chinese adolescent Internet users meeting the criteria for "addiction." It has been reported that addictions to video games are the fastest-growing forms of Internet addiction, especially in China, Taiwan, and Korea. Interestingly, the highest numbers come from a sample of Italian adolescents—36.7 percent reportedly showed signs of "problematic Internet use."

A study of more than thirteen thousand adolescents in seven European countries in 2014 found that 13.9 percent of the participants demonstrated what was described as *dysfunctional Internet behavior* due to compulsive and frequent use that resulted in problems at home, in school, or in general. In a breakdown of excessive usage, social-networking sites like Facebook gobbled up a lot of their online time, along with watching videos or movies, doing homework, downloading music, sending instant messages, and checking email. Boys were significantly more likely to be at risk for the more serious condition of Internet addictive behavior, with boys from Spain and Romania scoring the highest rates, and boys from Iceland the lowest. The more educated the parents, the less likely the adolescents were to show problems.

The study concluded that about 1 percent of adolescents exhibited Internet addictive behavior and an additional 12.7 percent were at risk. Together, this totaled 13.9 percent who could be said to demonstrate dysfunctional behavior. That means that more than one in ten of these adolescents are at risk.

Along with Kimberly Young, another pioneer in the field of addiction to technology is Dr. David Greenfield, a professor of psychiatry at the University of Connecticut School of Medicine and director of the Center for Internet and Technology Addiction. In 2014, in conjunction with AT&T, Greenfield conducted a telephone survey of one thousand subscribers and concluded that around 90 percent of Americans would "fall in the category of overusing, abusing, or misusing their devices." Greenfield, who is also the author of *Virtual Addiction: Help for Netheads, Cyber Freaks, and Those Who Love Them,* says the incidence of Internet addictive behavior among Americans is around 10 to 12 percent, according to his research.

What are people checking on their phones? The first-quarter results for Facebook in 2016 showed that its users spent an average of fifty minutes a day on the site, which is, according to a *New York Times* article, just a bit less time than most people spend each day eating or drinking.

In Greenfield's survey of phone use alone, 61 percent of respondents said they slept with their mobile phone turned on under their pillow—or on a nightstand next to the bed. More than half described feeling "uncomfortable" when they forget their mobile phone at home or in the car, travel somewhere and are unable to get service, or break the phone. Greenfield's research found that while 98 percent of respondents said that they are aware that texting while driving is dangerous, nearly 75 percent admitted having done it. This is effectively extreme risk-taking, the sort of lack of behavioral control that is usually associated with impulsive and compulsive behavior.

A few years ago, a research colleague of mine proposed to do a study assessing mobile phone addiction. He prepared the research proposal and set about recruiting participants. The idea was that all participants would hand over their mobile phones for a period of time, five or six days, while their levels of anxiety would be measured. Not one person approached was willing to participate in the mobile phone separation-anxiety project—which sort of proves the case.

So what can we do?

Internet addictive behavior expert Kimberly Young recommends three strategies:

1. Check your checking. Stop checking your device constantly.
2. Set time limits. Control your online behavior—and remember, kids will model their behavior on adults.
3. Disconnect to reconnect. Turn off devices at mealtimes—and reconnect with the family.

In other words, it's a revision of Timothy Leary's 1960s mantra, *Turn on, tune in, and drop out.*

Turn off, tune in, and reconnect.

Compulsive Shopping

There have been shopaholics forever, since the open-market takeout restaurants of the ancient Romans (yes, they had takeout in Pompeii). It is defined by compulsive, episodic purchasing of goods and is similar in many ways to gambling addiction. It is not recognized as a formal disorder due to insufficient evidence, but that would be cleared up quickly if a panel of experts studied a frantic sales line at midnight before Black Friday.

Shopping addiction is also known as compulsive shopping, compulsive spending, and compulsive buying. It is often trivialized in the media, and the so-called fashion victims described are invariably female. Like other problematic behavior, it can be easily amplified and escalated online.

Formerly, a tendency for compulsive shopping could be inhibited, and more easily self-regulated, by store hours, the need to transport oneself to a shopping site, not to mention the difficulty of carrying all those bags. A compulsive shopper, like anybody with a compulsion, has a lack of self-control. Now, because of technology, the obstacles to addictive shopping behavior have been removed. It is much harder for those with a tendency in this direction to resist.

Psychologist Elizabeth Hartney, an addiction expert, has studied compulsive shopping and explains that shopping online is particularly seductive to real-world shopping addicts because it appeals to many of the same motivations behind real-world shopping addiction, which are "the need to seek out variety in and information about products; to buy without being seen; to avoid social interactions while shopping; and to experience pleasure while shopping."

Recognize yourself?

Why do people feel compelled to buy things they don't need? And spend money they don't have?

The psychological explanations for compulsive shopping depend on the approach or school of thought. Traditionally, the behavior is believed to be triggered by a need to feel special or less lonely. Suffering from low self-esteem, the compulsive buyer is thought to be in a search for self—and looking for identity and stability in purchases, objects, or the social status that they feel is granted them once the new object is in their possession. Many suffer from associated disorders such as anxiety, depression, and poor impulse control.

It is also a form of addiction that is strongly encouraged by our consumer culture and the corporations that drive it. As Donald Black, a professor of psychiatry at the University of Iowa College of Medicine, has pointed out, "In America, shopping is embedded in our culture; so often, the impulsiveness comes out as excessive shopping."

Some experts have argued that compulsive shopping is a form of OCD. Just as mobile phone technology makes it harder for compulsive individuals to avoid checking emails, online shopping offers more enticements or "signals" to seduce the compulsive shopper—with intrusive advertising, pop-up coupons, pop-up sales, and many more images of goods to buy.

There are also new competitive, and patently more exciting, forms of shopping online—like the brilliantly designed eBay, which has a clock to tell you exactly how many seconds are left before the auction you are bidding in will be closing. And if you need another reminder, eBay will happily send an alert to your phone.

If you ask around for anecdotes, you'll find that there aren't too many eBay regulars who haven't gotten up in the middle of the night to watch the bidding close on an object they want—and they will tell you in great detail how exciting it was to try to beat the other bidders. This fun was lessened when robot services like BidRobot and Auction Sniper could be hired to do your bidding for you. That's when online auctions became a competition not just for goods but to see who has the better bot.

It's nice of eBay to offer a first-person blog about shopping addiction, otherwise known as "Retail Therapy Syndrome," on its website. The blogger reports that the number of online shopping addicts is estimated at 13 million, but her page has had only 8,700 views, a tiny fraction of the 162 million active users of eBay reported for the last quarter of 2015.

Whether or not her blog has helped many eBay addicts, it captures in a paragraph what multiple academic journals can only struggle to convey: "I feel a literal rush winning auctions and getting items at great prices. I *LOVE* not having to leave the comfort of my home where I can browse stores online at any hour of the day in my PJ's, and guess what? When you shop online you will never see a sign saying: 'No Shirt, No Shoes, NO SERVICE!' "

Email checking and excessive texting have overlaps with eBay addiction, because of the immediate rewards and the fever quality of the auction environment escalated by the cyber environment. As Kimberly Young writes in her blog, Netaddiction.com:

> In more serious cases, eBay addicts feel a sense of accomplishment when they discover they are the highest bidder and begin to bid on

items they don't need just to experience the rush of winning—sometimes to the point that they go into financial debt, take out a second mortgage, or even go into bankruptcy just to afford their online purchases. One client stole funds from her husband's 401k until he discovered her addiction. "He shut down my account and threatened to divorce me," she explained. "I was about to lose my marriage all because I couldn't stop myself from using eBay."

In evolutionary terms, I believe, the compulsive online shopper is engaged in accelerated "seeking." The cyber-shopper feels the rewards of foraging and finding, hunting and gathering. Like the other real-world addictive behaviors, there's a vicious cycle of online escalation: In order to try to feel better, the shopper experiences a high from online shopping and spending that is followed by a sense of disappointment and guilt, which can precipitate another cycle of impulsive shopping and spending in order to feel better again. According to Ruth Engs, Professor Emeritus of Applied Health Science at Indiana University, some people will take their purchases back because they feel guilty, which, once again, can trigger another shopping spree. Compulsive shopping is a way of self-medicating, but it leads to more stress, anger, isolation, disappointment in oneself, or depression. This is the classic roller coaster of addictive behavior.

As the shopper's debts grow, his or her behavior is often conducted secretly. Compulsive shopping is similar to other addictions: While alcoholics will hide their bottles, shopaholics will hide their purchases. When this level of shame is reached, and the purchases are either hidden or destroyed (before they can be discovered), the consequences can be devastating. Marriages, long-term relationships, and careers can be threatened or ruined. Further problems can include ruined credit, bankruptcy, and even in some cases suicide.

What are the warning signs? Rick Zehr, formerly of the Illinois Institute for Addiction Recovery at Proctor Hospital and now president of the Institute of Physical Medicine and Rehabilitation, offers the following list. Want to take the test? Answer yes or no to the behaviors you have engaged in.

1. Shopping or spending money as a result of feeling angry, depressed, anxious, or lonely.
2. Having arguments with others about one's shopping habits.
3. Feeling lost without credit cards—actually going into withdrawal without them.
4. Buying items on credit rather than with cash.
5. Describing a rush or a feeling of euphoria with spending.
6. Feeling guilty, ashamed, or embarrassed after a spending spree.
7. Lying about how much money was spent. For instance, owning up to buying something, but lying about how much it actually cost.
8. Thinking obsessively about money.
9. Spending a lot of time juggling accounts or bills to accommodate spending.

Here's the bad news. If you answered yes to four or more of these, according to Zehr, you may have a problem. And you might want to get some professional advice about it.

Gamer's Thrombosis

Compulsive gaming is by far the most studied cyberbehavior. It is most prevalent among young men, and studies report that compulsive gaming is often found together with mood or affective disorders, including depression, anxiety, and ADHD.

The other thing: It can be fatal.

Twenty-year-old Chris Staniforth of Sheffield, England, knew what his passion in life was: online gaming. In 2011, the bespectacled supergamer with brown hair and a kind face had just been admitted to Leicester University to study computer game design. While he owned a range of different consoles, and collected them, he had recently gotten a new Xbox and was playing a lot of *Halo,* a bestselling online military sci-fi shooter game—since the series was first launched in 2001, more than 60 million copies have been sold—in which players battle invaders from outer space.

"Chris lived for his Xbox," said his father, David Staniforth. "When he got into a game, he could play it for hours and hours on end, sometimes twelve hours in a stretch."

After a day of marathon gaming, Chris woke up in the middle of the night with a strange feeling in his chest but was able to go back to sleep. The next day, during a visit to a job placement center, he reached down to pick up a pack of gum that he'd dropped and felt a jolt, then a spasm, and fell to the ground. A friend who was with him called an ambulance, but it was too late. A local coroner confirmed that Chris had died of pulmonary embolism. A blood clot that had formed in the veins of his lower body—deep vein thrombosis, or DVT—had traveled to his lungs, blocked his arteries, and killed him.

DVT is caused by inactivity, by sitting still for long periods of time. In the past, those at risk were passengers of overseas flights and bedridden hospital patients. Americans might remember the NBC White House correspondent David Bloom, who died in 2003 in Iraq after spending hours immobile in a tank while embedded with the U.S. infantry. An estimated three hundred thousand to six hundred thousand people develop DVT each year in the United States, and between 10 and 30 percent of all patients die within thirty days. So-called extreme gamers—the 4 percent of the total U.S. gaming population who average 48.5 hours a week in front of a screen—are at risk for DVT, which has led the U.S. media to call this life-threatening condition "gamer's thrombosis."

In Asia, where marathon gaming is more prevalent, competitive gamers or extreme gamers gravitate to popular public arcades or gaming halls called Internet cafés that are open twenty-four hours a day in some countries. In the West, we probably think of an Internet café as a quiet coffeehouse that offers Wi-Fi or an Internet hookup, with maybe a few spare computers to borrow. In the East, the term increasingly describes a phenomenon that has taken over India, China, South Korea, Taiwan, and the Philippines, where the old computer game and pinball arcades have been replaced by thousands of gaming centers—sometimes called "PC bangs" or "LAN centers"—where rows and rows of chairs and consoles provide gamers with a PC with a LAN high-speed connection for as little as a dollar an hour. There, surrounded by dozens of fellow gamers, a gaming enthusiast can binge-game, usually going online to meet dozens of other gamers on a MMORPG (massive multi-player online role-playing game) like *League of Legends, World of Warcraft, Counter-Strike,* or *StarCraft.* Evenings, after dinner, the cafés are traditionally filled with competitive gamers, usually over eighteen years old, who are welcome to pay and play all night.

In this exciting setting, you can imagine the additional incentives to keep playing—surrounded by other gaming enthusiasts in the room, and cheered on by your gaming cohorts online. The question is, where is the gamer? Are they sitting in the real-world environment of the café, or are they immersed or even lost in cyberspace? A number of fatalities point to a break with reality.

In 2012, an eighteen-year-old boy in Taiwan, a gamer who went by the name "Chuang," booked a private room in an Internet café and holed up there without eating or sleeping for forty hours to play in a marathon session of *Diablo III,* then died of what is suspected to be a fatal blood clot. The same year, also in Taiwan, Chen Rong-yu, twenty-three, played *League of Legends* for a day straight and was found dead in a PC bang with his hands stretched out for the keyboard and mouse. Gamers continued to enter and leave the room, oblivious to the dead young man in their midst.

There are more than twelve thousand Internet cafés (and counting) now in South Korea, where professional gaming is now a multimillion-dollar industry. But the cafés have come under scrutiny, and appear to be the epicenter of dark tales and addiction. A spike of concern was provoked when an unemployed married couple became so hooked on a role-playing game called *Prius Online,* a virtual world where they were caring for a virtual baby named Anima, that they neglected to feed or care for their actual baby, who'd been born prematurely and was unhealthy. When the real baby died of malnutrition at three months old, the couple was reportedly gaming at an Internet café. (This became the subject of an HBO documentary, *Love Child,* in 2014.)

Other disturbing tales have surfaced in the media—almost too awful to believe. In 2010, a twenty-two-year-old Korean man was charged with killing his mother when she nagged him to stop gaming. He left her corpse in order to return to the café and keep playing. Another man, gaming at a café in the city of Taegu, played *StarCraft* continuously for fifty hours and died of a heart attack.

When reading stories like these, of such passionate engagement, you almost have to remind yourself, *This is a made-up game in a virtual world.* What causes this kind of dedication, persistence, and devotion?

In one significant study of excessive gamers, brain imaging was done while subjects described their urge to play games online and recalled various gaming experiences that were provoked by pictures. The results showed that their cravings had a similar neurobiological pattern as the cravings for drugs in drug addicts and the cravings for alcohol in alcoholics.

The brain doesn't lie. This study and others show a biochemical component in compulsive online activity, which means the behavior could qualify as a "true addiction" and "disorder," using the clinical definition. In 2007, when the American Psychiatric Association (APA) began preparations to update DSM, one of the subjects discussed at length was compulsive Internet gaming.

Internet Gaming Disorder

When the APA identifies a new clinical disorder, it's a big deal in my world—and a game-changing event for mental health practitioners, patients, and their families. It is a decision meant to stand the test of time, an acknowledgment that a behavior has been consistently proved by studies to be serious and is understood well enough for criteria for diagnosis to be established and proper medical intervention and treatment options suggested. Once "legit" in the eyes of the APA, a disorder receives more widespread attention in the media and journals of medicine and can qualify for health insurance coverage.

The impact on private and public mental health practitioners, who aren't on the front lines of research, is tremendous. They turn to—and trust—DSM to give them strong and clear guidelines about how to diagnose a disorder and care for real patients in their offices and clinics.

A work group of twelve members of the APA and twenty outside advisors with expertise in related specialties of substance use and addiction was assembled and met over a period of five years to discuss compulsive Internet gaming, as well as the more general term, *Internet addictive behavior.* They looked at studies and literature on potential nonsubstance-addictive-related behaviors—including gambling, Internet gaming, Internet use, shopping, exercise, and work. (Excessive eating and excessive sexual behavior were discussed with the eating disorders and sexual disorders work groups.) The result?

In the case of Internet gaming disorder, the evidence in 250 studies done in countries around the world shows a persistent and escalating problem of youth, adolescents, and young adults, mostly male, who demonstrate symptoms of addiction-like behavior while playing online games. When DSM-5 was released in 2013, it was noted that Internet Gaming Disorder is most common in male adolescents 12 to 20 years of age. According to studies it is thought that Internet Gaming Disorder is more prevalent in Asian countries than in North America and Europe.

The APA suggested that these four components were essential for behavior to be described as Internet addiction and recommended further study:

1. Excessive Internet use, often associated with a loss of sense of time or a neglect of basic drives.
2. Withdrawal, including feelings of anger, tension, and/or depression when the computer is inaccessible.
3. Tolerance, including the need for better computer equipment, more software, or more hours of use.
4. Adverse consequences, including arguing, lying, poor school or vocational achievement, social isolation, and fatigue.

In China, one of the first countries in the world to define overuse of the Internet as a clinical condition, and perhaps where the problem of compulsive gaming is perceived as more serious and prevalent, more action has been taken. The government developed treatment centers to cure teens of "Internet addiction." In South Korea, the government found an escalating number of "addicted" teens who were compromising on sleep, schoolwork, and real-world friends in order to play online games. In 2011, a "Shutdown Law" was enacted, prohibiting those sixteen years old and younger from logging online between midnight and 6:00 a.m. The following year, an amendment prevented under-eighteens from playing online between midnight and 6:00 a.m. unless specifically allowed by their parents or guardians. As of 2015, there were five hundred "Internet addiction" inpatient units in hospitals and treatment centers in Korea, as well as prevention programs in place in every single school system. The most interesting recent development is in Taiwan, where legislation was passed in 2015 that allows for parents to be fined up to 50,000 Taiwanese dollars ($1,595 U.S.) if their child's use of electronic devices "exceeds a reasonable time." It is unclear how "reasonable" is defined or how the government plans to monitor it.

The climate seems to be 180 degrees different in the United States, where a discussion about making "e-Sports," or gaming, an official college athletic sport has begun. Kimberly Young argues that this seems irresponsible in the face of study results that have found gaming associated with many problems. Take, for example, the 2004 story of thirteen-year-old Zhang Xiaoyi in China, who played *World of Warcraft* for thirty-six hours straight—what's called a gaming marathon—and afterward jumped out the window of a tall building, having left behind a note saying that he wanted to join the heroes of the game. If you play a game for thirty-six hours straight, any game, you are likely to be mentally and physically exhausted to the point of being delusional.

But the problem of gaming isn't simply a matter of excessive playing. Young writes on her blog that at her addiction center, "we see addicted gamers who are more than twice as likely to have ADD/ADHD, get into more physical fights, and have health problems caused by long hours of game play (e.g., hand and wrist pain, poor hygiene, irregular eating habits). Many need treatment to improve their academic performance and return to normal functioning."

She continues: "[T]here should be great concern about American colleges deeming video games as sport. It is important that we first understand the impact of these games on our youth. While video games can be fun and entertaining, I continue to hear from families who are struggling because of a child's gaming habits. What may seem like a competitive sport could be masking a deeper problem."

ADHD + Technology

In the late 1960s and early 1970s, a series of studies on impulsivity and delayed gratification were conducted by psychologist Walter Mischel at Stanford University that are now referred to as the Stanford Marshmallow Experiments. A child was offered a choice between one small reward provided immediately or two small rewards if they waited for a short period, approximately fifteen minutes, during which they were left alone in the room. The reward was sometimes a marshmallow. Other times it was a pretzel or a cookie.

Some children had no difficulty waiting out the fifteen minutes in order to get the double reward. Some children were incapable of that. In follow-up studies that went on for decades, the researchers found that children who were able to wait longer for the enhanced reward tended to have better life outcomes, as measured by higher SAT scores, higher educational attainment, even better body mass index (showing, I suppose, restraint with food consumption), and other life measures.

Clinical studies—and observations we can all make about people in our lives, or even ourselves—show us that impulsive individuals have a harder time with self-control. They can lose themselves more easily in an activity and in the moment. They tend to be comfortable with, or seek out, risk. They have more difficulty stopping an activity that is rewarding to them but that may be self-destructive or dangerous.

We know from one study that children and teens diagnosed with some behavioral disorders are more likely to become addicted to the Internet. What is interesting is the gender component. It breaks down like this:

- Boys diagnosed with ADHD or hostility are more likely to become addicted to the Internet.
- Girls diagnosed with depression or social phobia are more likely to develop an addiction to the Internet.

What does this say to me? Let's remember that correlation does not imply causation. There are predispositions and tendencies at play, and there could be more reasons for this than impulsivity. For example, a boy who has been diagnosed with ADHD may also be socially disconnected from his peers. These boys can be physically rough, impulsive, and nonengaging. Misunderstood by their classmates and unappreciated on the playground, they may gravitate to the computer.

And children with ADHD are highly spirited and restless and have a hard time focusing, and therefore could be put in front of a computer screen more often. If given a highly exciting, stimulating, and ever-changing game to watch and interact with, they can become engrossed. Parents may even say things like, "Gosh, he's a different child when he's playing video games. He's finally sitting still!" For a parent, seeing a hyperactive child sitting immobile (for once) in front of a computer screen might be a welcome sight.

Some parents may believe they need to find ways to "tire out" their ADHD kids. But I would argue that the child already may be exhausted—and needs to do less, not more. Creating quiet time may be more important, not giving him more activity or stimulation. A scent diffuser or a lavender- or orange-scented candle burning may help—these scents are known for their therapeutic effects. In fact, any child would benefit from a quieting experience, like sitting in a room with low lighting, just talking with a parent or listening to a soothing audiobook.

Designed to Addict

The gaming industry will never forget the video game crash of 1983, known in Japan as "the Atari shock." Revenues were peaking at around $3.2 billion in 1983 and then fell to around $100 million by 1985—a drop of almost 97 percent. A number of explanations were offered at the time, from the end of the teenage fad for gaming to market saturation, and then there was a disappointing launch of *E.T. the Extra-Terrestrial* adventure video game, developed and published by Atari. Whatever the reason, the crash was brutal—and one the gaming industry was determined not to repeat.

Now far bigger than the U.S. film industry, and bigger than network and cable television, the gaming industry brings in $66 billion annually. A big game like Vivendi Universal's *World of Warcraft* (often called *WoW*) can spend years in development and cost as much as $60 to $100 million to produce. With that commitment of money, talent, and time, it's no surprise these games are so ingeniously compelling, or that game design teams have a sense of pride about how "addictive" their creations are, and about the level of audience engagement and investment they engender.

Fun failure and other hooks are built into successful games in order to keep players playing. A well-designed game is supposed to cast a spell over players—and make playing irresistible and compulsive. Anticipating the release of a new game fuels the frenzy. Even before a much-anticipated new game is launched, fans begin preordering it online, and if they fail to get their new game, there's frantic participation in the so-called secondary markets that sell prized content online at auctions.

Of course, that's the point, isn't it? Entertainment is supposed to be entertaining. Hollywood hopes to make movies and TV shows that are engrossing. The book publishing industry hopes to release books that "you can't put down." When an audience finds a form of entertainment that is super-gratifying, you hear about it in the signs of compulsive behavior—sleeping in line on the street to buy Bruce Springsteen concert tickets, binge-watching the entire series of *Making a Murderer* over one weekend, waiting up until midnight for the release of the latest *Star Wars* movie.

When Disney invests millions of dollars to install a new ride at one of its theme parks, the goal is to offer something new and newly exciting—and a ride that people will come back, again and again, to experience. The point of these thrilling rides is that you are supposed to feel like you are in danger when, in fact, you are not. What about an entertainment that seems safe but may be dangerous?

Andrea Phillips, a game designer with Transmedia, told the technology news site Thenextweb.com that she believes that different compulsive behaviors are "produced" by different games:

> FPSes [first-person shooter games] and MMORPGs tend to maximize length of play session; whereas Zynga-style social and casual games maximize number of sessions [via encouraging] a return to the game as often as possible. I do find the Zynga-style social, mo-

> bile games more evil, if you will, just because many of these games are very close to compulsion loops and nothing else.

It's interesting that she uses the word *evil*—and by that, I assume she means *manipulative*. In forensic psychology, we don't use this word. *Evil* is a primitive construct, a return to the days of fearing the unknown. As a profiler, I think it can tell you more about the person using the word than it does about whatever they are attempting to describe.

Phillips describes "compulsion loops" as tasks that are repeatedly required, done over and over by the gamer in order to reach the next level of play. Game designers intentionally employ these loops, fully aware that they are derived from behavioral experiments of classical conditioning. They are similar to the hooks that encourage excessive gambling, using positive reinforcement to create addictive patterns. When the designer talks about players, it sounds a little like she's talking about lab rats that are being trained.

Phillips explains:

> It's that tension of knowing you might get the treat, but not knowing exactly when, that keeps you playing. The player develops an unshakeable faith, after a while, that THIS will be the time I hit it big.
>
> THIS is the time it will all pay off, no matter how many times it hasn't so far. Just one more turn. One more minute. But it's really never just one more. . . . For the most part, I steer clear of multiplayer situations, MMOs [massive multiplayer online games], and so on because I just can't trust myself. With narrative games with an ending, I know I'll binge-play them, so to avoid the fallout of missed sleep and deadlines, I don't even start a game like that unless I have a good solid week with no serious commitments.

A high engagement level is a sign of a successful game—and "good" game design. An example would be *World of Warcraft,* where gamers are in an immersive environment designed to keep them playing as long as possible—and that means extreme time commitments of uninterrupted hours spent engrossed and immobile—because there is no designed "end" to the game. No resolution, no grand finale.

Do game production companies ever grapple with the fact that they are producing games that are consumed by young audiences and encourage excessive playing? What is their ethical responsibility to the well-being of the public? Food manufacturers have to list caloric content and ingredients on their packaging in the United States. Fast-food companies like McDonald's and others have relented, under intense scrutiny and long campaigns against heart disease, high blood pressure, and obesity, to offer meals that are healthier. But so far, the regulations governing games are similar to the ones governing Hollywood films: They are given a rating for content, which measures its acceptability in terms of sex, violence, and other mature themes.

So what about player well-being?

Should there be a measure of irresistibility, as in *Most Likely to Cause Immobility and Addictive Behavior*?

I believe that there is an element of human exploitation in these game designs. And as much as they claim to be superrealistic, they don't involve players in a way that a real soccer game or real basketball game would. In a real game, as it goes into a second half or final quarter, a player's energy dissipates. The game becomes actually less difficult, due to the tiredness of most of the other players. The pace slows down. In the final quarter of a basketball game, the players are wiping off their sweat with towels and they are sometimes out of breath. This is a natural progression, and something we don't even notice as spectators because it is so universal, a natural law of game playing.

The opposite occurs in most online games. The longer you are playing, the more difficult the game becomes. You enter new worlds, new levels, where a higher skill level is required in order to succeed. As you play, you become more fatigued—and possibly hungry or sleep-deprived. You're barely awake and conscious. And your willpower is low and your judgment impaired. To stop playing at this point actually requires too much energy.

I call this *unnatural game design*. Why not make levels easier instead of harder as you progress? Because millions of dollars have been invested, and teams of brilliant designers have been employed, to keep you playing, and playing, and playing. In multiplayer games, the game goes on without you—and often your gaming cohorts online will try to convince you not to stop. *You can't quit now! We need you!* If you have been marathoning for hours, as gamers are wont to do, each hour invested in a game makes it harder to walk away. Your success can feel like a matter of life and death.

And, as we know, it can be.

Gaming Freak-Outs

A camera records a young German boy sitting at a desktop computer. He has short blond hair and baby fat in his cheeks. My guess is he's about thirteen or fourteen years old. In the short video that's been uploaded on YouTube, he is seen avidly playing an online game. His eyes are glued to the screen. Something upsets him—the game doesn't go the way he hoped. He begins rocking back and forth in his chair. Things seem to get worse, and he begins hyperventilating, pounding his fists on his keyboard. "Think positive! Think positive!" he chants to himself, while rocking. Suddenly he laughs maniacally, as if he's totally out of control.

"What? What's this?" he begins screaming at the screen. The game is going terribly, and he can't get control of himself. He screams louder and louder. He is foaming at the mouth; his breath becomes shallow, his voice hoarse. He begins pounding even more violently on the keyboard—and keys begin to fly behind him, and other plastic pieces. He chants: "I don't need help, *I don't need help!*"

But doesn't he?

If you don't believe that young people have trouble quitting an addictive online game after they've logged eight hours of play or more, then go on YouTube and check out "Gaming Freak Outs." These are viral videos of kids—usually teenage boys—becoming hysterical while playing an online game. These videos are popular for teen viewing because they are extreme and therefore entertaining. But for me, as a cyberpsychologist, it's pretty hard to find them funny.

So engrossed in playing a game, and so invested in success and continuation, a gamer becomes inordinately distressed when he makes a mistake, or is interrupted by a person in the real world (often a younger sibling), or is told that he must stop playing, usually by a parent.

What happens next is a tantrum and meltdown and breakdown rolled into one. Gamers scream and cry, explode into obscenities, throw their Wii remotes into a TV screen—and sometimes break it. There are many videos of small children, under nine, shouting words they shouldn't know yet while being pulled away from violent shooting games like *Call of Duty,* which they shouldn't be playing.

While teens and other gamers love to watch freak-outs and cheer and laugh and share their favorites, what they are seeing is really a kind of dissociative episode, because these young people no longer have a handle on reality or even a memory of it. (As many as 41 percent of the participants of a 2009 study said they played video and computer games to escape reality.) The only world they recognize in that moment is the gaming world. It is an all-consuming, compulsive behavior. The only thing they care about is succeeding at the game and satisfying their craving to play.

But it's actually even worse than that. These freak-outs, cruelly recorded by people who are supposedly "loved ones" and "friends" of the gamer, are then posted online. The gamer who is experiencing a terrifying breakdown—which is truly not funny—will be humiliated again and again, possibly for the rest of his life, as fellow gamers and other strangers around the globe watch his freak-out and cry with laughter.

The humiliation is also eternal. That's the online *escalation* piece of this new behavior, something that technology makes possible. Prior to the Internet, the proverbial "sore loser" of a kickball game on an outdoor playground might dissolve in tears but would be soothed, and the scene would be over in a matter of minutes. Kids being kids, that playground tantrum might not be forgotten immediately, but it would likely pass into folklore, and the memory of it, even for the sore loser, would fade in time. There would be no actual record of it. Online, a freak-out can last forever.

Locus of Control

Depression is another psychiatric disorder commonly found alongside the compulsive disorders, both in the real world and online. In a study of 370 Greek adolescents, the main predictor for compulsive Internet use was the amount of time spent on the Internet, with a significant comorbidity—or concurrence—with depression and "low locus of control."

Locus of control is a psychology construct developed by Julian B. Rotter in 1954 that describes a person's perspective on life and whether he or she attributes successes and failures to individual strength and perseverance—or to a twist of fate. Locus of control is about one's belief in personal control. *Locus* is Latin for "place" or "location." Your locus of control is said to be *internal* (sometimes described as "strong") if you take responsibility for your life and outcomes and believe you are driving events. Your locus is internal if you believe hard work and preparation are the foundations for success rather than good luck. An apt analogy taken from a children's book would be *The Little Engine That Could,* the small train that gets over the tall mountain by cheerleading itself to a successful outcome. So-called pop psychology and self-help books would refer to this as the "power of positive thinking." Call it what you want, but I would never disparage anything that helps people overcome adversity and live happier, more rewarding lives.

Your locus of control is *external* (sometimes described as "low") if you believe your decisions and life are controlled by environmental factors that you cannot influence, because everything is dictated by chance or luck—or somebody else is running the show. You feel powerless, in other words. An individual with a "low locus of control," as was found in the study of the 370 Greek boys, believes he has very little if any control over his life.

What is the relationship between locus of control and interaction with technology? Imagine the adolescent who is struggling with an Internet compulsion and feels powerless to effect a change. Imagine how easily this behavior, which is known to escalate, could become unmanageable and destructive. Whether you call it "Internet addictive behavior" or "Internet gaming disorder," the many studies done by various disciplines from psychology to neuroscience show that adolescents who compulsively use the Internet are at increased risk of quite a few negative social, behavioral, and health consequences—including poor academic performance, disorganized daily life, and difficult personal relationships. In other studies, adolescents with Internet addictive behaviors were more likely to demonstrate aggressive behavior, significant among adolescents in their middle school years. Online chatting, adult pornography site viewing, online gaming, and online gambling were all associated with aggressive behavior.

The effect of attitude on outcome has been proven time and again. It's like that great line attributed to Henry Ford: "Whether you think you can, or think you can't, you're right." Powerlessness escalates and feeds on itself. If you feel out of control, the Internet may seem like a better environment—and a way of altering locus of control. Highly responsive, the Internet and the devices we use to access it can give users a feeling of power. Dr. Martin Seligman, a cognitive behavior psychologist at the University of Pennsylvania and the author of *Learned Helplessness* and *Flourish,* describes the super-responsiveness of technology and gadgets as offering the user a false sense of control. "These technological gadgets are more responsive than anything we have ever seen on this planet," Seligman writes. "The problem is that the outcomes over which they give us such exquisite control may be trivial. . . ."

> [T]hey promise more than they deliver. Rather than allowing us to get to the substance of life in a more efficient way, they have become the substance itself, crowding other matters—murkier and less responsive to be sure—out of the scene.

Seligman's work in the past decade has been to devise new approaches to retraining attitudes in the field of positive psychology. He believes that the epidemic of depression among the young is actually a "disorder of personal control."

"How else can people judge their competence," Seligman writes, "except by ascertaining the control they have?"

Freud would call this *repetition compulsion.* People engage in behaviors that give them a feeling of control, but paradoxically they cannot control those behaviors. This is classic addiction.

In short, technology has the upper hand against a young adolescent—providing the illusion of control while undermining it.

Pleasuring Yourself to Death

Since the Internet's very first image ever transmitted—an image of a sexy pinup girl—cyberspace has been a fertile ground for sexual and sexualizing content. And just think where the human urge to see sexually available women and men, or naked people engaging in sex acts, has gone in the past two decades. The predominance of porn sites online is unarguable—they now receive more visitors each month than Netflix, Amazon, and Twitter combined—and represents 30 percent of the total Internet industry. In 2015, the cellphone pornography business was estimated to have reached $2.8 billion.

Compulsive cybersex is a significant component in Internet addictive behavior for many men and women, according to *The American Journal of Drug and Alcohol Abuse*. Therapists report a growing number of patients addicted to sex online, with the standard problems associated with addictive behavior. In some cases the behavior was initiated by early exposure to pornography and conditioned by reinforcement. Other compulsive users have underlying trauma, depression, or addiction. But both men and women with cybersex compulsions show behavior that is described as "maladaptive."

The subject of sex addiction has become as mainstream as it gets, represented by a growing number of celebrity poster boys like mouthy British comedian Russell Brand (who has also admitted to being addicted to heroin, cocaine, and booze) and *Californication* star David Duchovny, a brainy Princeton grad, no less. Not that long ago, there was no doubt of its being a diagnosable disorder. In 1987, DSM-3-R referred to sexual addiction as a sexual disorder "Not Otherwise Specified" (NOS). The term *NOS* is used to describe disorders of sexual functioning that are not classifiable in any of the specific DSM categories, and historically it was the description most commonly used for patients identified as "sexual addicts." (One of the examples given for this disorder is "Distress about a pattern of repeated sexual relationships involving a succession of lovers who are experienced by the individual only as things to be used.")

Just as the topic of sex addiction was becoming more recognized and discussed, it became controversial within the APA. Why? How? In preparation for the fifth edition of DSM, Dr. Martin Kafka, a Harvard Medical School professor and psychiatrist, reviewed the entire body of scientific research and literature on the subject of *hypersexual disorder,* concluded that it very definitely existed, and prepared a proposed diagnosis for inclusion as an official disorder.

His findings in 2009:

> Hypersexual Disorder is a sexual desire disorder characterized by an increased frequency and intensity of sexually motivated fantasies, arousal, urges, and enacted behavior in association with an impulsivity component—a maladaptive behavioral response with adverse consequences. Hypersexual Disorder can be associated with vulnerability to dysphoric affects and the use of sexual behavior in response to dysphoric affects and/or life stressors associated with such affects. . . . Hypersexual Disorder is associated with increased time engaging in sexual fantasies and behaviors (sexual preoccupation/sexual obsession) and a significant degree of volitional impairment or "loss of control" characterized as disinhibition, impulsivity, compulsivity, or behavioral addiction. . . . [Hypersexual Disorder] can be accompanied by both clinically significant personal distress and social and medical morbidity.

And yet the APA chose to disregard Kafka's presentation of the facts and decided to exclude sex addiction from DSM-5. This decision is still shrouded in mystery. I can't help but wonder what arguments experts made to one another to justify this omission. Perhaps the academy feels there is not enough scientific evidence proving that sex can become an addiction? Dr. Kafka clearly states that "the number of cases of Hypersexual Disorder reported in the peer reviewed journals greatly exceeds the number of cases of some of the codified paraphilic disorders such as Fetishism and Frotteurism."

Given the effect technology has on sexual behavior, this omission feels both very odd and very much behind the times. Studies show that sexually compulsive individuals are more attracted to the Internet because their behavior, which often requires secrecy, is easier to satisfy there due to what Al Cooper, of the San Jose Marital and Sexuality

Center, referred to as "the Triple A Engine." The A's stand for *anonymity, accessibility,* and *affordability*.

Anonymous, your secret is safe. With easy access, you can find an almost limitless supply of partners while you are at home, at work, at school, in a cyber café, or on your mobile phone. The affordability of just a few dollars per month means that almost anyone can access sexually based websites from a computer somewhere.

Excessive use of online pornography is regarded by some psychologists as a manifestation of both Internet addictive behavior and sex addiction. And of the reported sex addictions, pornography addiction is the most commonly reported, particularly among younger people. The definition of a "Phase One sex addict" is an individual who is addicted to sex that doesn't involve others. This could describe untold numbers of adolescents today. The behavior is reinforced and ingrained in the same way other compulsive behaviors are—and develops very much like a drug addiction. After an initially rewarding experience with pornography, the urges become more frequent and more powerful. These connections can become so strong that simply sitting down at a computer elicits a sexual response.

Take the story of University of Pennsylvania psychologist Dr. Mary Anne Layden, whose patient, a young male who was so addicted to Internet pornography that he missed an appointment for a job interview. He literally couldn't stop his behavior. Layden tells this story in a podcast about the growing problem of men—young and mature, students as well as fathers and husbands—who have lost all interest in their lives beyond their compulsive behavior. Brain imagery of online porn addicts shows a similar pattern as it does with other addictions. The craving is real—and destroys lives.

This reminds me of a study done in 1954 by researchers Peter Milner and James Olds in which the pleasure centers of rats' brains were stimulated by electrodes that led the rats to forgo all efforts to live regular lives. The rats had to press a lever to be stimulated, then exhausted themselves pressing it over and over again. They stopped sleeping. They stopped eating or drinking water. They literally pleasured themselves to death.

Again, gender plays a role. Psychologist Philip Zimbardo, of the legendary Stanford prison experiment, argues that young men are becoming hooked on arousal via online pornography and the results are catastrophic for both schoolwork and relationships with others. In Zimbardo's notable TED Talk in 2011 ("Why Are Boys Struggling?") and his recent book, *Man (Dis)Connected: How Technology Has Sabotaged What It Means to Be Male,* he describes the damage to boys from excessive Internet use, excessive gaming, and pornography consumption. Zimbardo categorizes these as *arousal addictions,* and says, with "drug addiction, you want more. Arousal addiction you want different—you need novelty."

I cannot help but feel that this resonates strongly with seeking—specifically in an online context. The average boy watches fifty pornography videos a week, according to Zimbardo's estimates, which partly explains why pornography is the fastest-growing industry in America, increasing $15 billion annually. By the time an average male is twenty-one years old, he has played ten thousand hours of video games (two-thirds of that in isolation). Both gaming and porn, Zimbardo argues, have effectively caused a digital rewiring of boys' brains to need novelty, excitement, and constant arousal, which means they can be totally out of sync in traditional classrooms and romantic relationships.

It's time to have a discussion about online pornography addictive behavior, or arousal addiction. When people talk about it, they discuss men, teenagers, and boys all in the same spectrum. But in reality, you are talking about very different things. A grown man who develops an interest in online pornography is one thing. But for a young boy who does the same, during critical developmental phases, the problem behavior can manifest in a different way and therefore become more complex and much harder to treat.

Oculus Rift—VR Headsets

Around the corner—and perhaps by the time this book is published—is mass distribution of Oculus Rift, a virtual-reality headset for 3-D gaming. With a helmet that will literally encase your head, eliminating the possibility that you can see the real world, or even a hint of real sunlight, the user will be psychologically and physically immersed in the cyber environment, which will only amplify and escalate the levels of engagement, investment, and addiction. The technology was purchased and is under development by Facebook, which will undoubtedly do a bang-up job of marketing it too. If you think the adolescent gamers are having trouble giving up their remotes now, just wait until you try to get them to take off their helmets.

I feel certain that this amazing technology could be used for better purposes. Along those lines, I have recently published an extensive protocol to use virtual-reality head-mounted display units (VR HMDUs) for the treatment of post-traumatic stress disorder (PTSD), involving a full range of sensory stimulation. On a positive note, my research in this area also makes me wonder if this technology isn't a potentially powerful way to relax and engage a child with behavioral difficulties. I am imagining a calming immersive VR experience of swimming with dolphins or floating down a river on a raft.

My underlying belief is that the problem is not technology—but what it is being used for. For the moment, it seems designed to always overstimulate, when just as easily it could be soothing.

Abstinence or Adaptation?

Technology offers enormous advantages in terms of accessibility and choice. We can shop online for food and have it delivered conveniently to our door. We can shop for presents and have them delivered to someone else's door. We can purchase everything from clothing to electronics with ease. The problem occurs when this behavior escalates and spins out of control. As it stands now, technology offers the optimum environment for this to happen.

While academics argue whether there is sufficient evidence to describe any of these behaviors using standard addiction criteria, the problems continue to evolve. There is no denying them. Rather than debate their existence, it would be more productive if we could talk about treatment. Abstinence, one of the proven methods for recovery from addiction, is not possible. Giving up alcohol, cocaine, or cigarettes may be an option in the recovery process for those addictions, but giving up technology is not. Which is why I believe we need to rethink our approach to the concept of Internet addictive behaviors.

The Internet is not going away. We are moving to a place where we don't have a choice but to engage—if you want to study, have a job, do research, be informed, access your healthcare benefits, or pay your bills. Technology has become as natural as the air we breathe, as necessary to twenty-first-century survival as the water that replenishes our bodies. It has become part of our environment. Therefore the challenge could be an evolutionary one. Rather than talking just about addiction, we could discuss *adaptation.*

The best way for an individual to adapt effectively to a new environment is to be informed—know the new environment well, study it, and be aware of the dangers and pitfalls. The Internet could be the most beguiling and enticing creation of humankind. Our evolutionary instincts, such as seeking, make it hard to resist. Not to mention the hooks, signals, alerts, prompts, fun failure, and heart-stoppers. To adapt intelligently, we need to know ourselves well and learn how all things cyber affect our behavior.

For an individual who has behavioral control issues or attention deficits, or who is predisposed toward compulsive behavior or is simply higher on the impulsivity scale, the cyber environment may be problematic. These people are more vulnerable online. It can cause trouble for them the way a racetrack or casino causes trouble for a compulsive gambler. Take caution if you are given to impulsivity.

To me, whether we are clinically "addicted" may have already become an arbitrary argument. Are we addicted to running water and hot showers? Are we addicted to the personal freedom granted by the automobile? A Darwinian approach may be better. *Maladaptive behavior* is a term used to describe attitudes, emotions, responses, and patterns of thought that result in negative outcomes (for example, biting your nails when you're nervous). When presented with a situation, a maladaptive approach seems to render good results, but these are temporary and can lead to greater problems.

If we consider our struggles with cyberspace, it may be shortsighted to get bogged down in the addiction debate. If technology is a substance that we need—and is central to our survival going forward—then we need to learn to live with it, but on our own terms. Once we realize the specific triggers that facilitate cyber-seeking frenzies, the ones that drive dopamine surges and fuel impulsive, compulsive, or addictive behavior—whether it is online shopping, gaming, or pornography—technology could be designed to be more compatible with its users. Or we could learn to adapt and become more resilient, more restrained and disciplined, more focused and less compulsive.

We are still at the beginning of an unimaginable shift in how we live. Let's give ourselves a break. If you have a problem with technology, perhaps you're not addicted, just *cyber maladapted*. And the good news: There are things you can do about that.

CHAPTER 3

Cyber Babies

Not long ago I was sitting on a train going from Dublin to Galway. A mother and a baby came to sit across the aisle. I had some work to do, some reading, but as we headed west toward the Atlantic Ocean, the countryside of Ireland was so beautiful, I took a break and just enjoyed the view out the windows. Looking at the ever-changing cloud formations passing over the dramatic Irish landscape always relaxes me.

The roads and fields of Galway are lined by spectacular stone walls. The land is stony, and over the centuries, each time a field was cleared, more stones were dug up and more walls and cottages were built. Looking through the window, I thought about all the labor that went into building these walls—the moving of each stone and the artistry of the masons, some of whom also worked on the ancient towers, churches, and cathedrals of Ireland.

Stonemasons loved to work on cathedrals because their stones held the spires aloft. They reached for the sky, for the heavens. And by building the cathedrals, the stonemasons believed they were reaching for something better. Each generation needs to reach for something better—to build things, to create a legacy. As the train jostled and chugged along, I began to ponder the works of civilization being created now. What fields are we clearing and what is being built? Where are our cathedral spires?

Across the aisle, the mother settled in and began feeding her baby. In a wonderful display of dexterity, she held the bottle in one hand and clutched a mobile phone in the other. Her head was bent to look at her screen. We're all busy these days, and it's hardly my business to judge how a young mum feeds her baby. Out of the corner of my eye, though, I observed her with a researcher's curiosity. *Ethnography* is the immersive study of people and cultural phenomena, when the researcher is embedded in the social group being studied. As a cyberpsychologist, I am living in a continual ethnographic study. Hardly an hour goes by when I don't notice how people are interacting with technology.

Ten or fifteen minutes passed. The mother looked exclusively at her phone while the baby fed. The baby was gazing foggily upward, as babies do, and looking adoringly at the mother's jaw, as the mother continued to gaze adoringly at her device. For half an hour, as the feeding went on, the mother did not make eye contact with the infant or once pull her attention from the screen of her phone.

Forget the lovely view of the Irish countryside, by this time I was riveted to the human scene unfolding across the aisle from me. It was a good thing I was wearing sunglasses.

My mind filled with questions. I couldn't help but wonder how many millions of moms and dads around the world were no longer looking directly into the eyes of their babies while they were feeding or talking to them. Obviously, they might look sometimes, but what if that direct contact was in fact one-half or one-quarter as much as the days when my generation was raised? How will this seemingly small behavioral shift play out over time?

Would a generation of babies be impacted?

Could it change the human race?

Parents usually ask me this question: What is a safe and healthy age for a baby to be introduced to screens? They mean anything from iPads, tablets, and mobile phones to televisions. Without a doubt, this is a very important question. But first, I ask parents to think about this question:

What is the right age to introduce your baby to *your* mobile phone use?

Face Time with Your Baby

Did you know it's really important to *look at your baby's face*? Feeding and diapering alone aren't enough. A hug and a quick kiss aren't enough. Babies need to be talked to, tickled, massaged, and played with. And they need your eye contact. There is no study of early childhood development that doesn't support this.

By experiencing your facial expressions—your calm acceptance of them, your love and attention, even your occasional groggy irritation—they thrive and develop. This is how *emotional attachment style* is learned. A baby's emotional template, or attachment style, is created (or "neurologically coded") by the baby's earliest experiences with parents and caregivers. When a good secure attachment is formed by consistent interactions between baby and parents, the template for future emotional connections is more secure. And a secure attachment pattern will give a baby a much better chance at becoming a confident and self-possessed individual who is able to easily interact with others. An individual's attachment style affects everything from how we form friendships to how we choose a life partner to how our relationships may end.

The science of love is fascinating. In the 1930s, Harry F. Harlow, an American psychologist, looked at infant-mother bonds and the impact of maternal separation by studying baby rhesus monkeys whose mothers had died. He noticed that infants raised in a nursery became very different adults from those raised with mothers. These orphaned monkeys were slightly strange, reclusive, and very attached to the soft blankets in their cages. When these blankets (they were actually cloth baby diapers) were taken away once a day and cleaned, the orphaned babies became super-agitated and upset, often curling into fetal balls and sucking their thumbs. This led Harlow to suspect that physical contact and the role of the primary caregiver were much more important than previously believed. At the time, institutional nurseries and hospitals did not provide physical contact for human babies; it was considered unnecessary, even unhealthy.

Harlow did an experiment allowing the orphaned baby monkeys to choose between two surrogate mothers. One was a wire-framed "mother" that had a milk bottle to feed the baby. The other "mother" was made of soft terry cloth—but had no milk bottle. All the monkey babies preferred their soft terry-cloth mother with no milk, and spent their days cuddling and hugging them, visiting the "bottle mother" only for feeding.

The babies chose tactile love over food in every instance. In a second experiment, Harlow scared the monkeys with a frightening sound or other stimulus in their cages—a banging noise and a robot-type contraption with flashing eyes and moving arms. Panicked and under stress, the babies ran quickly to their soft moms for comfort, clinging to the terry cloth and calming down.

Some baby monkeys were not given soft terry-cloth mothers; instead they were given the wire-frame milk-dispensing machines, and these babies showed much less emotional resilience. When they were frightened by loud, unexpected sounds, they were unable to cope or be comforted. They threw themselves on the ground, clutched themselves, or rocked back and forth, shrieking. This proved that infant love and bonding is not simply a matter of feeding but an important and crucial psychological resource that made the monkeys emotionally stronger.

Like the monkeys, a human being's attachment-style pattern is formed early, but not set in stone. Our emotional template and other templates are continually being updated throughout our lives. The human brain is often described as "plastic," which means that it is able to change—physically, functionally, and chemically. And for a human baby, a terry-cloth mother who provides only tactile comfort is not enough. Many experiments in the past century have shown the catastrophic effects of sensory and social deprivation during certain critical periods in early childhood, and the subsequent effects on later development. And healthy attachment or "connectivity" patterns have been proven to be important to a child's intellectual growth and progress.

How does bonding work?

A mother and her child need to be paying attention to each other. They need to engage and connect. It cannot be simply one-way. It isn't just about your baby bonding with you. *Eye contact is also about you bonding with your baby.* In terms of evolutionary theory, infants are soft and cuddly, cute and round, and completely adorable because their appearance lures adults to look at them—attention that they critically need. This is true of baby animals in almost any species. Think of bear cubs, puppies, kittens. (Their popularity on social media networks alone should prove this!) They are soft, round, and supercute, which to date has given them an edge in the competition for attention. If babies weren't so cute, they might not learn to talk, walk—or do hundreds of other things that interaction with their human caregivers teaches them.

When these bonds and connectivity patterns are not formed properly—something that's usually seen in neglectful or abusive homes, or in environments like large orphanages and other institutions where children receive infrequent interaction and don't have the opportunity for tactile stimulation and exploration—a child may even fail to develop the neural pathways necessary for learning. When the deprivation is extreme enough, a child may fail to thrive, may never learn to bond with others. Or learn to love.

We know this because of the examples of children who were raised in isolation or in the wild. There are at least nine famous cases of *feral children* over the past century. And while these stories are sad, the information about child development gleaned by scientists has been invaluable. In each case, the children were accidentally separated from their families, or ran away, causing them to grow up without human contact during their formative years. In several cases, children were raised with packs of wild dogs or wolves, like Mowgli in *The Jungle Book*. How this impacted the child depended on the age when all human contact was removed. But in several cases, children who were raised with dogs or wolves walked on all fours—and barked. A Russian boy who was raised in a room with dozens of birds (his negligent mother treated him like a pet) was rescued at the age of seven by social workers. He couldn't speak, and when he was agitated he chirped and flapped his arms.

There are almost no incidents of a wild child developing into a fully functioning adult (or at least a conventional version of one), because there are windows in the formative years when very specific skills need to be learned. When those developmental windows close, a child may be developmentally or emotionally crippled for life.

For parents raising their child in a normal household, the simple idea that a baby needs a mother's eye contact—a small but significant aspect to child rearing—probably didn't need to be emphasized until recently. Until sixty years ago, apart from listening to music or the radio, what else was there to do while you fed your baby?

But in the media-saturated environment of the average household, this has changed. And given the size, portability, and interactivity of phones and tablets, an Internet connection and a wireless device now vie with an infant's need for one-on-one interaction on trains, park benches, and even sitting at home on the couch. When fifty-five caregivers with children were observed in fast-food restaurants by researchers for a 2014 study in *Pediatrics* journal, a vast majority of them (forty caregivers) used mobile devices during the meal and some (sixteen caregivers) used their devices continuously, their attention "directed primarily at the device" and not the children. These devices are so compelling that they can overwhelm basic human instincts. Together with the fact that the workday for most adults has become round-the-clock—another change facilitated by technology—parents come home to their young families still distracted by work questions and interruptions. At the office, cyber-slacking is what happens when you check your Facebook wall when you should be knee-deep in a financial-forecast Excel spreadsheet. But the reverse can be true at home. Instead of spending time with their children, many parents find themselves still distracted by their devices. (Remember that statistic from chapter 2 that mobile phone checking actually increases after work hours?) I assume that in many households, older siblings of these babies are likewise distracted, often entertained by screen games and fun apps rather than occupying themselves with the new baby (as all bored older siblings once did).

Who is the real loser in this new scenario?

The baby.

It scares me to think that this could be the first time you are reading about the importance of face time with your baby (and I don't mean the app kind of FaceTime). Perhaps it has been drowned out by louder debates. Society has passionately discussed the nutritional values of breast-feeding and the best age and method for potty training. Governments regulate the designs of car seats and mandate the use of them while banning lead paint and flammable sleepwear and bedding. It's hard to pick up a marble or a Lego piece without seeing a printed warning label about choking hazards for children under two.

Of course the primary thing is to keep your baby safe. And obviously, you'll find no argument from me about that. Regulations against the marketing of unsafe toys save thousands of lives each year and result in many fewer children with brain damage due to asphyxiation or poisoning by toxic chemicals. But it might be time to reconsider other less immediate and less obvious dangers.

Am I saying that unless you stare constantly at your baby it will become feral? Obviously not. But I am saying that if the average baby born in 2016 receives significantly less one-on-one interaction, less eye contact, than the average baby born in 1990, there will be an effect or change.

Like what kind of effect?

Let me take a few guesses, the kind that academics like me aren't really encouraged to make—speculation is not considered "good science," but in some cases it is very necessary. Over time, people could become less able to interact face-to-face, less sociable. People could become less likely to form deep bonds with others, less able to feel or give love, and therefore less likely to form lasting relationships, families, and communities. Some could find physical contact with other human beings problematic and even unwelcome. There could be a domino effect. Subsequent generations could be raised with even less attention, less love—or none at all. While it's true that humans are gregarious by nature, and a search for connection is a basic human instinct and a survival skill, it does not happen magically and on its own. Real-world face time is required. This small and simple thing, millions of babies around the world getting less eye contact and less one-on-one attention, could result in an evolutionary blip.

Yes, I said it. *Evolutionary blip.*

Less eye contact could change the course of human civilization.

But so far, there aren't campaigns to alert parents to the dangers of their wandering attention span. Nobody seems to be even talking about this issue, this real risk, except those with an interest in cyberpsychology. If the lobbying group for babies (who don't really contribute to the economy, do they?) were as strong and rich and persistent as the ones for Big Pharma, real-estate agents, retirees, the technology industry, and commercial banks, someday there might be writing on the screen of all mobile phones that says:

> *Warning: Not Looking at Your Baby Could Cause Significant Developmental Delays.*

Infants: Show Me the Science

The baby on the train to Galway was small, less than a few months old, a time in a baby's life that is referred to by some developmental scientists as the "fourth trimester." At three months, a baby is still quite fetal. During this remarkable period, his or her brain will grow about 20 percent in three months.

Experiences in the world are what keeps a baby's brain growing—and what keeps a baby developing properly. When a baby is born, each cell of the brain has around 2,500 synapses—the connections that allow the brain to pass along signals. In the next three years, that number grows to about 15,000 per brain cell, when the brain creates 700 to 1,000 new neural connections every second. Synapse formation for key developmental functions such as hearing, language, and cognition peak during this time, making this window in a young child's life extremely crucial for the development of higher-level functions.

This was the thinking behind the Baby Einstein products, which were first marketed in 1997 by a former teacher and stay-at-home mom, Julie Aigner-Clark. She and her husband, William Clark, invested $18,000 of their own savings to produce the first product, a video they called *Baby Einstein,* meant for infants and children under two. It showed toys and cartoons and other visuals interspersed with sounds and music, stories, numbers, and words in several languages.

Just four years later, the Baby Einstein franchise was bringing in $25 million a year, and several companies had invested to become part owners, including Disney.

The premise?

Stimulate your baby's brain and you can increase a baby's intelligence, or even create a baby genius.

Except . . . what about the science?

Too much stimulation is not necessarily a good thing.

Visual acuity, as it's known in the field of child development, is acquired in the first two years of life, if a baby is raised in normal real-world conditions. This window of time is crucial in the creation of properly functioning eyesight. Similar to the way language skills should be acquired before five years of age, the same goes with depth perception and binocular vision, which is a factor in hand-eye coordination, balance, and fine motor skills. At birth an infant's visual acuity ranges from 20/200 to 20/400 (the higher the number, the worse the eyesight) and improves rapidly in the first years of life, further evidence of the incredible changes that happen in the infant brain during this period. By the time infants reach two years of age, most of them have miraculously achieved 20/20 vision.

Perceptual development is a true neurobiological wonder—and a product of nature and nurture. It doesn't happen without an environment. As soon as they are born, babies begin scanning the world around them and looking for meaningful patterns.

During the first two months of life, their eyes focus on edge detection and shapes, a process that has inspired work in the science of *computer vision* (or *image understanding*). At three months, a baby's focus shifts to *internal features* of an object, or the features within a shape. There's a part of the brain in both children and adults that is dedicated to *face processing,* or facial recognition, something that we know from brain imaging studies. Babies demonstrate an ability to prefer their mother's face from the earliest hours of life, and by two to three months old they show a preference for the internal features of her face, particularly her eyes.

This means, as much as adults are hardwired to find babies irresistible to look at—for the survival of the species—a baby will prefer to look at its mother's face and eyes over other things. This is how development and learning begin.

Can an animated app, an avatar, or a 3-D cartoon video replace this, re-create it, or override human nature?

In 2006, nearly a decade after the initial launch of its products, a complaint was filed against the Baby Einstein company, which had become a booming multi-million-dollar-a-year global brand, for making false claims. Eventually several studies backed up this complaint, alleging that young children who viewed the videos regularly for one month, with or without parents, showed no greater understanding of words from the program than children who had never seen it.

Even more troubling, a research team of developmental experts coordinated by the University of Washington studied infants between eight and sixteen months who were exposed to videos and DVDs such as those sold by Baby Einstein and Brainy Baby and found this exposure to be strongly associated with lower scores on a standard language development test. In other words, the study claimed that these videos could delay speech. The interpretation of the study findings were softened later, when the founders of Baby Einstein sued the University of Washington and a second team of experts was brought in to interpret the findings—and determined that the videos had minimal to no impact on infant development. Negative impact was declared "not sufficiently proven." Disney, which by then owned the Baby Einstein brand, changed the wording of the product claims on its merchandise and offered a refund to anyone who had purchased the DVDs for language enhancement purposes and hadn't seen results.

Let's take the Baby Einstein debate out of the courts for a second. And let's put aside whether the University of Washington study was interpreted correctly the first time. This is the salient question: What is really known about brain development and the acquisition of language (and other cognitive skills) in infants?

The evidence is irrefutable: The best way to help a baby learn to talk or develop any other cognitive skill is through live interaction with another human being. Time and time again videos and television shows have been shown to be ineffective in learning prior to the age of two. Most significant, a study of one thousand infants found that babies who watched more than two hours of DVDs per day performed worse on language assessments than babies who did not watch DVDs. For each hour of watching a DVD, babies knew six to eight words fewer than babies who did not watch DVDs. Still, some formats, some shows, and some ways of delivering educational information to young children have been shown to be more effective. These seem to be quieter shows, calmer formats with only one story line—the television show *Blue's Clues,* for instance, and *Teletubbies.*

Children who are taught by their parents and caregivers, people with whom they have an emotional bond, demonstrate the most improvement. Researchers speculate that this is probably because very young children learn through gestures and interactive communication with adults. In other words, babies learn best from humans, not machines.

So why do these "early learning" products continue to sell well? There is a never-ending parade of them, as any trip to iTunes, Amazon, or a baby store will show you. A recent and mind-boggling example of this was a 2013 Fisher-Price product called the Apptivity Seat, designed to "grow" with your child. For a newborn, it was a bouncy seat. For a toddler, it became a walker. In both cases, it was a sedentary entertainment contraption with an "overhead pivoting case" that held an iPad or tablet over the baby or toddler's face. Just $74.59 on Amazon, it came with free-to-download apps that were advertised as "developmental," "soothing," and "early learning."

In the sales photograph shown on Amazon, an infant is pictured with an iPad only a baby's-arm-length from its face. (This is probably because the optimal viewing distance for a newborn is six to twelve inches.) This way, even before the baby has the motor skills to lift the device or the neck strength to turn away, the poor thing is trapped by technology. It's less of an Apptivity Seat and more of a *captivity* seat.

One special feature was advertised: "Locks your iPad device securely inside case to protect from dribbles and drool." That's very thoughtful. The iPad stays clean, but in the meantime, who is bothering to protect Baby?

Quickly dubbed the "worst toy of the year" by consumer groups, the Apptivity Seat was soon discontinued by its manufacturer. The apps, however, are still available online—and can be used with the Apptivity Gym, which positions a device above your baby's head while he or she lies on a mat.

Here's what troubles me. Let's start with the fact that a mobile phone or any wireless device is considered by many public health experts to be a risk and possible carcinogen for newborns, due to the unknown effects of radiation exposure on their fragile, developing systems. We also don't know how looking at the screen of a tablet might impact an infant's eyesight development. A very interesting series of studies from 1958 (that led to a Nobel Prize) was carried out on kittens, revealing that visual experiences at birth, when the brain has a high degree of plasticity, have permanent and irreversible effects. These studies are really important in terms of the impact of visual stimuli such as exposure to digital screens, as they demonstrate that sensory input is central to visual development in newborns.

Next, there's this concern: If a baby spends too much time simply being cyber-stimulated and not connecting in the real world—with real people, real pets, real toys, and real objects—it could impair other important pre-academic skills such as empathy, social abilities, and problem-solving. These are things primarily learned by exploring the natural environment and using the imagination to spend time in unstructured, creative play.

Simply moving in the world—movement for movement's sake—is not just good for the heart, lungs, circulatory system, and everything else in a human being's body; for a child it is essential to participate in sensorimotor activities. Moving in the world is what eventually leads to knowing how to pick up a block, climb a tree, run downhill, and build a sand castle. These visual-motor skills have also been shown to be important to later success in math and science.

The Apptivity Seat was described as promoting "discovering" and "learning" (not to mention being "soothing"). If you're like me, you may wonder who is regulating the educational value of videos, DVDs, and tablet and phone apps for infants. How can these companies make such positive claims when, in fact, it is possible that overexposure to these products may do more actual harm than good?

The answer is disturbing on many levels. It seems that as long as no hard science exists to *disprove* their claims, companies feel free to say almost anything they can get away with. It saddens me to suspect that profit and pure greed is the driving force, not the welfare of children, because until conclusive studies can be done, we are in the dark. And until then, there is really no way to scientifically evaluate these new tech devices and products, or even know if the delivery vehicle for them, the touchscreen and its computer-generated depth perspectives, isn't having a harmful effect on visual perception and eyesight. Only studies done over time will reveal the overall impact. By that time, a critical developmental phase for an entire generation of young children may be over.

But let's start with some things that are proven, and have been confirmed by short-term studies:

- Babies and young children do not truly understand what they are seeing on a screen until they are approximately two years old; therefore the experience cannot enhance knowledge, understanding, or cognitive skills.
- When a screen is on, a baby is less likely to play on its own—exploring the physical world—which is how real learning takes place.
- When a screen is on, parents tend to talk to their child less, which is detrimental to a baby's language learning. More screen time also means less eye contact and facial reading.

Now I'll deliver the final blow. Prepare yourself. More than fifteen years ago, in 1999, the esteemed body of the American Academy of Pediatrics recommended against screen use, including television, for children under two.

I'm curious. Had you heard about that? At first, the AAP offered this as a "suggestion" because what was known about media consumption among infants then was inconclusive. But in the ensuing fifteen years, as more studies have been done, the recommendation is now based on findings. In 2011, the AAP issued a science-driven policy statement discouraging media use by children under two.

That's right. You read that correctly. Or did you miss it?

No TV for babies. No apps with funny cartoons on a parent's or babysitter's mobile phone. The AAP believes these things could potentially have a negative influence on a child's development. But even so, there's been an explosion in electronic media marketed directly at infants aged one to eighteen months, a multimillion-dollar industry selling computer games for very young children—some as young as nine months—and a launching of entire television networks that specifically target children as young as twelve months. And who could forget the Apptivity Seat, labeled as suitable for newborns?

So why isn't this important AAP cautionary information provided on the marketing materials and packages of baby apps and tablet software targeted at them? I wish I had an answer for you.

Toddlers and Tablets

The tablet is now ubiquitous as a "toy" for toddlers (two to three years old)—and parents often marvel at the swiftness with which their child learns to swipe a touchscreen. So simple, so easy, it seems almost intuitive. These devices, as well as mobile phones, have become a game-changer in terms of screen time for children, lowering the age of the user considerably. Unlike a desktop or laptop, a tablet can be used by any child who is old enough to point a digit.

The fundamental problem, I believe, is the modern perception (or misconception) that children need to be kept busy and occupied at all times. And, like the thinking behind the Apptivity Seat, designed to keep a wandering child confined while enrapt by technology, giving a child a tablet is a convenient way for parents or caregivers to grab a few minutes, or an hour, for themselves.

What's the harm, right?

And besides, what about all those other parents giving their children these little handy screens? Millions of people can't be wrong, can they?

But they are.

This is a field I've researched in depth—and in 2015 published a review paper, "Cyber Babies: The Impact of Emerging Technology on the Developing Infant." It's hard to know where to start, as I begin unpacking all my concerns about cyber babies using devices. To begin with, there could be physical risk. As stated above, we still don't know how the levels of radiation exposure of a mobile phone or tablet can impact children, who are certainly more vulnerable in this regard than adults. A 2013 report in the *Journal of Microscopy and Ultrastructure* warns that fetuses are most vulnerable to radio frequency energy exposure (radiation) and mothers should not carry mobile phones in their clothing.

Next, I have to raise the problem of credibility of the "educational" claims made by the app developers. Thousands of these have already been created and marketed for toddlers and children under five.

Somewhere along the line, a misinterpretation of neuroscience has led parents to believe that all stimulation for a child is good stimulation. They believe, wrongly, that a young brain must be kept constantly challenged and engaged. It's as if parents fear their toddler will become bored with real life, which I guess means life without a screen. And fearing a toddler tantrum may be another reason that tablets and mobile phones are pulled out of handbags and totes by parents and caregivers to placate and pacify young children. I have seen articles by psychologists referring to these as "shut-up toys."

Even if these devices in themselves are not proven to be harmful, there is significant harm simply in the lack of time spent doing things in the real world that are known to be important for development. It has been shown repeatedly that at least sixty minutes per day of *unstructured play*—when children entertain themselves, either alone or with another child and without adult or technological interference—is essential. This is when a child uses imagination and creativity, when he or she practices decision-making and problem-solving. This unstructured play is what helps children to develop early math concepts, such as shape, size, sorting, order, and simple counting. At the same time, they develop fine motor skills and hand-eye coordination.

This is why many developmental experts believe that the best toys are the ones with the fewest rules. By playing, the toddler has something to learn about the world.

What about a child who spends the bulk of his or her playtime with an interactive app, in which objects explode, appear, reappear, and don't play by the rules of the physical world? It's possible this habit could interfere with the reaching of *object permanence.* This milestone was discovered when the renowned developmental psychologist Jean Piaget set out to study how a child's knowledge of the world changes with age. He argued that a child is an active participant in the development of knowledge and asserted that cognitive functioning develops in stages. He outlined four stages of intellectual or cognitive development; the one from birth to eighteen months is *sensorimotor development.* At a certain stage in a toddler's intellectual growth, he or she will realize that a toy continues to exist even if it has been removed—or is out of sight. This understanding by the child is called *object permanence.* Some would argue that magical thinking contributes to creativity, but the toy is a physical object—it is matter—and, according to the fundamental laws of physics, it cannot simply disappear. If this developmental milestone of logic is not reached in an appropriate way, during a specific window of time before the age of three, then we have no idea of the consequences in terms of cognitive development.

Given the ethical constraints of doing potentially harmful scientific research on infants, or any children, it is possible that we may never know conclusively the impact of interactive screens until it's too late. Much of what we do know about classical conditioning and early childhood development was produced by an array of pretty horrific experiments, done prior to the ethical debate about proper and humane treatment of lab subjects. Sadly, a string of American psychologists did with babies and small children what Ivan Pavlov, the famous Russian psychologist, did to dogs.

Perhaps the most troubling and groundbreaking work was done by behaviorist John B. Watson and his graduate student Rosalie Rayner at Johns Hopkins University in the 1920s. They experimented with a nine-month-old baby, "Little Albert," the son of a wet nurse who lived and worked on the campus of the university hospital. The wet nurse was paid one dollar for Little Albert's participation in a series of experiments in which Watson set out to prove that a fearful response to almost anything could be instilled in a child with the proper conditioning.

Poor Little Albert, a trusting and good-natured chubby blond cherub, was put in a sterile laboratory environment and subjected to endless sounds and scary sights as Watson studied the effects of conditioning. The child was shown a series of stimuli including a white rat, a rabbit, a monkey, masks, and burning newspapers. Initially, the child showed no fear of these things.

Then, in order to instill a fear of rats in Little Albert, when he was next shown a white rat, Watson hit a metal pipe with a hammer to make a loud, disturbing sound. Little Albert began to cry, as any baby would. The white rat and the loud, disturbing sound were paired repeatedly until, eventually, Little Albert started sobbing at the sight of the white rat alone.

Fear—and therefore other emotional responses—had been successfully conditioned. If you go online and google "Watson" and "Little Albert," you can see lots of pictures of the poor little guy crying.

B. F. Skinner was another behaviorist who did groundbreaking work in what he called "the Skinner box," a controlled environment where animals could be closely observed while they were stimulated by lights, sounds, projected images, and even electrical shocks—and also where specific behavior could be reinforced by food and water. That was all well and good, and the animal-rights lobby and PETA weren't around to complain or throw paint on him. It was a different story when Skinner put his own daughter Deborah in a box he designed, a continuously warm, controlled environment that was an oversized metal "air crib" with three walls and a glass window, where she slept and played for the first two years of her life. The point Skinner was trying to make with his air crib (which he designed to be manufactured and marketed) was really about convenience to parents—while creating a warm, safe environment for a baby. Deborah and three hundred other children were raised in such beds, and reported no ill effects. But even so, in 1945 there was an uproar when a photograph of Skinner's daughter in what seemed like a glass cage was published in *Ladies' Home Journal*.

Would anybody cause a fuss if *Ladies' Home Journal* ran a photograph of a two-year-old child playing with a tablet or mobile phone? Has this vision already become a norm? Is there anything we can do to correct that?

Preschoolers + Literacy

Of course, not all technology is ruining childhood or negatively affecting development. We live in a cyber world, and kids need to learn to navigate it—but, more important, parents need better guidelines as to the effects of cyber. I have no doubt that e-books can be useful in promoting vocabulary development and reading comprehension in older children, aged four or five. There is promising research suggesting that they may increase early literacy skills by providing practice with letters, phonics, and word recognition. And they may be more engaging for young children because of all the bells and whistles, the animation, funny voices, sound effects, touchscreen, highlighted text, and embedded games. A roaring 72 percent of iTunes's top-selling "educational" apps are designed for preschoolers and elementary school children. There's a lot of money to be made in technology, and all of us, including parents, are too easily brainwashed by the limited science behind new products. Just go to an online app store and search the word "kids" and you'll get more than sixty thousand results. The so-called educational app media market has grown into a billion-dollar industry. This is *scientism,* or the exaggerated belief in the power of science, at its worst.

But exactly the very things that can make these apps exciting and engaging have been shown to distract children from actual learning. Extraneous e-book enhancements interfere with comprehension or a child's ability to simply follow a story. So while your child seems to be sitting still and enjoying the "book," he or she is actually just stimulated but learning nothing.

It is well known that young children learn less from video than from live interactions, something developmental psychologists call the *video deficit effect.* Often what can be achieved by live learning in one instance takes repeated attempts in a video. When young children are spoken to directly by parents or caregivers, they absorb and learn language very quickly. But they do not learn language effectively from the television or from observing conversations between adults because their attention is not directed enough. They are distracted by other stimulation. Along the same lines, studies have shown that having the television on at home all day as background noise causes similar language delays. Children can't hear human speech properly when there is constant sound in the house.

The promotions for apps always refer to them as "interactive," a buzzword for parents, probably because they've heard or read somewhere that children learn best through "interaction" rather than passive observation. And indeed, using a highly interactive app is an entirely different dynamic for a child than watching television, but I would argue that it is not a better one. The interactive aspects of a tablet could make it even more distracting and far too stimulating for a young child. Not all screen time is created equal.

What's the best route? What age to start, and what screen?

There are a growing number of voices and opinions on this matter. My concern is primarily with the millions of parents who have been left adrift while new norms have been established with the help of marketers and the clever tech industry rather than based on behavioral or social science. (I find it ironic that there's been such a proliferation of "science" content for kids and yet apparently little actual science about developmental impact.) Without governance or cyber ethics regarding the claims of the app manufacturers, the nursery is a free-for-all, making the conundrum more daunting for a new parent. How can we expect them to parse out these new devices, new apps, and new technology, and make the best decision for their families?

On one end of the spectrum, there's the toy industry, which will happily thrust the iPad into the face of the newborn infant. On the other end of the spectrum, there are some very conservative views, as expressed by Cris Rowan, a pediatric occupational therapist, child development expert, and founder of Moving to Learn, who recommends limited to no screen use for children:

- Only one hour of television per day for children ages three to five.
- No handheld devices or video games recommended before the age of thirteen, and a restriction on video games to thirty minutes per day for thirteen- to eighteen-year-olds.

According to Rowan, as many as one in three children now enter school developmentally delayed, negatively impacting literacy and academic achievement. The recommendations of Moving to Learn may seem harsh, but the signs of trouble are becoming more prevalent, as schoolteachers and primary school administrators are beginning to report a higher incidence of developmental delay in children entering kindergarten. In Britain, an escalation of problems associated with pervasive tablet use among preschool-age children has been reported by the Association of Teachers and Lecturers, including developmental delays in attention span, fine motor skills and dexterity, speaking, and socialization—as well as an increase in aggressive and antisocial behavior, obesity, and tiredness.

British teachers report a growing number of young children who are beginning school with an expert ability at swiping a screen but not enough dexterity to pick up and play with building blocks. One gathering of teachers in Manchester called for help with "tablet addiction." And a teacher in Northern Ireland described young students who were allowed to play computer games excessively before bed arriving at class the next day with what you might call a "digital hangover," and attention spans "so limited that they might as well not be there."

What happens when a child plays on a tablet but never has a chance to look at a real book?

Sadly, you don't have to go very far for the answer. You can see a one-year-old girl on YouTube, in a video that's had 4.5 million views.

A print magazine is given to the child. She puts the magazine in her lap. With her fingers, she expertly makes the touchscreen flicking motion on the printed magazine page and is stumped when the image doesn't change. Then she tries to make photographs larger by pulling apart her index finger and thumb, another touchscreen cue. But the concept of turning a page is alien to her.

She proves the point that a digital native may excel in front of a screen but will be disadvantaged in the physical world and perhaps not even recognize it as her domain.

Technology will provide us, inadvertently, with more such experiments, I fear. But there is no reason to give up and always expect the worst. A positive use of technology to study child development occurred recently when MIT researcher Deb Roy, a professor of linguistics who studies how children learn language, recorded every moment of his son's first three years with webcams placed around his home in order to study how he learned to speak.

Harkening back to the early days of conditioning experiments, Roy showed how tight feedback loops between the child and his father, mother, and nanny had effectively taught him to speak. (It wasn't from watching Baby Einstein videos.) Roy's popular TED Talk, "The Birth of a Word," shows in one minute how his child learned to speak one word, "water." To do this, ninety thousand hours of video were aggregated and edited by algorithms into short video clips showing the evolution of the sound "ga-ga" to the word "water." His work continues, and Roy has designed a device called Play Lamp, a less intrusive recording device currently being used in a pilot study of autism.

Sleep + Electronic Media

The art of falling asleep is a critical life skill. Pediatric sleep experts offer lots of indispensable advice for establishing good sleeping patterns in early life. (Some of these aren't so bad for adults, either.) Keep the lights down, play calm music, keep activity and excitement before bedtime to a minimum. There should be a bedtime routine of sorts, which will prepare the baby (or you) physically and by classic conditioning to fall asleep. These are cues, and they really work.

Watching television has been proven not to be a sleep inducer or a good pre-bedtime routine. Even worse is the exciting, interactive app. As much as tablets are given to children to pacify them, even the ones sold as "soothing" are still stimulating.

Screen use can cause sleep disturbances, in both children and adults. This seems to be true of all "light-emitting devices." The U.S. National Institutes of Health reports findings that watching a screen before bed—whether it is a computer, gaming console, tablet, mobile phone, or television—is associated with insomnia and subsequent daytime sleepiness. And computer use in teens and young adults—playing, surfing, or reading—causes sleeplessness, as does mobile phone use for the same purposes.

Why? The brightness of the screen can confuse the circadian rhythms.

It could be that people come to associate the screens with highly interactive pursuits and activities. If your child is used to playing exciting, superstimulating games on a tablet, then the child has an association with the device that it will be stimulating—and no amount of "soothing" apps may override that conditioning. Personally, I use an app on my devices that adjusts the color of my screen display for the time of day: warm at night, brighter, like sunlight, in the morning.

The fundamental problem is, once again, the modern perception that children need to be kept busy and occupied at all times. And then we have to find a technological answer for their overstimulation. Sherry Turkle, a psychologist and MIT professor of the social studies of science and technology and the author of *Alone Together,* a much-talked-about book that describes how technology is changing family life, explains this fear of having a "bored" child:

> Learning about solitude and being alone is the bedrock of early development, and you don't want your kids to miss out on that because you're pacifying them with a device.

Pacifying or overstimulating? Sleepiness, like boredom, is a natural state and doesn't require special devices and interventions, except the simple solution: going to bed. But it's worth noting that children and adults sometimes behave differently as a result of sleepiness. Adults usually become foggy and listless when tired. Some children overcompensate and become more energetic and even hyper. A sleepy child can be moody, emotionally explosive, agitated, or aggressive. In one study involving 2,463 children ages six to fifteen, children with sleep problems were more likely to be inattentive, hyperactive, and impulsive, and to display oppositional behaviors.

For this reason, sleep deprivation is sometimes confused with ADHD in children. According to the National Sleep Foundation, more than two-thirds of children in the United States experience one or more sleep problems at least a few nights a week. And children who have been diagnosed with ADHD have been shown to be more vulnerable to sleep deprivation, which can profoundly impact ADHD symptoms. One study found that treating sleep problems alone may be enough to eliminate attention and hyperactivity issues for some children.

The Rise of ADHD

Some early-learning experts believe there is a connection between the rise of ADHD and screen use in children. It is now the most prevalent psychiatric illness of children and teenagers in America. Is it a coincidence that neuroscience research shows that just as the iPad, iPhone, and other digital screens began saturating households in America, the number of ADHD diagnoses in children increased? We cannot imply causation; a relationship between two variables does not mean one causes the other. But I don't believe in coincidence. In forensics we say, "There is no such thing as coincidence, only actionable intelligence." This phenomenon requires urgent and informed investigation.

The number of young people being treated with medication for ADHD grows every year. And, quite shockingly, more than ten thousand toddlers, ages two and three years old, are among the children reportedly taking ADHD drugs, even though prescribing these falls outside any established pediatric guidelines.

Dr. Dimitri Christakis, a medical doctor and professor at the University of Washington, is a prominent voice in the conversation about the connection between attention disorders and interactive screen use. While looking after his own two-month-old son, Christakis realized that jumpy images on the screen were engaging the baby's natural reflex to react to a change in his environment. Subsequent research by Christakis and his colleagues found an association between children under three who watch more than two hours of television a day and a difficulty with attention. This doesn't necessarily mean that television causes the attention difficulties. As I have mentioned earlier, it can mean that children with behavioral problems may be put in front of a TV more often.

Further studies found that attention difficulties were more precisely linked to program content. While educational programs had no effect, both nonviolent and violent entertainment were associated with subsequent attention problems. Specific cartoons and fast-paced media were linked to attention difficulties. This led Christakis to theorize that overstimulation of the brain of a child might be harmful to his or her development. The style of some programming available on television, with quick editing cuts and quickly changing images, engage a baby's *orienting response,* the reflex that fixes attention to strange sights or sounds. According to Christakis, this reflex is what keeps a baby's attention focused on the screen, which can lead to an overstimulation of the developing brain.

These studies were only talking about TV, which is practically a media dinosaur when compared with the interactivity of digital devices, apps, and interactive games. Talk about hyperstimulation! As Christakis has said, one of the strengths of the iPad is that it is interactive, which allows a child to progress at his or her own pace and is therefore potentially better for educational purposes. But in fact, interactivity may be the problem.

The term *interactive* is a misnomer. As Dr. Leonard Oestreicher, a family physician and member of the Society for the Study of Autism Spectrum Disorder and Social-Communication, points out, there is no social activity possible while an infant or a young child is playing with an "interactive" device. "There is no eye contact, turn taking, true voices, or opportunities for joint attention," he writes. "These images cannot react to the child's social bids such as baby smiles, laughs, or babbling."

Christakis and his researchers found that the more television a baby or toddler watches, the more likely they are to have attention problems by the time they reach age seven, which is the average age at which children with ADHD are diagnosed. Another group of researchers found that prolonged screen use—either television or video games—may be connected with a risk of developing attention and learning problems, and associated with negative educational outcomes in the long term. It was shown to negatively affect executive functioning and cause attention deficit, cognitive delays, impaired learning, increased impulsivity, and decreased ability to self-regulate (that is, to avoid having tantrums).

Of course, there are lots of explanations for a rise in the ADHD diagnosis. The theories about the cause of the escalation continue to propagate. Could it be a diet of preservatives or too much refined sugar? Could it be too much homework? A lack of exercise? Or simply an increase in detection rates? But many experts believe that early screen exposure and the media-saturated home environment of today may be, if not the cause, a contributing factor in the rise of ADHD. To be sure, the pharmaceutical industry has taken full advantage and encouraged more diagnoses. Drugs are a mainstay of treatment. And are now, it turns out, handed out to toddlers.

Think about this for a moment:

Distracted by technology = reduced eye contact
A steady diet of technology = tantrum
Tantrum + more technology to pacify = escalation
Escalation of problem = drugs

Yes, these are big leaps. But they are plausible.

Good science often echoes good sense. And I think we all know what it's like to feel overstimulated—and be unable to concentrate. In one study of kindergarten children, two classrooms were used: one with plain, undecorated walls and another with highly engaging and colorfully decorated walls. Researchers found that students did better in the dull environment. The visually "busy" classroom impacted the students' ability to maintain focus on instruction and learn. "The decorated classroom led to greater time off task than the sparse classroom, and greater time off task in turn led to reduced learning."

Can the appetite for overstimulation be created in young children—similar to the way a consistent diet of sugary snacks has been shown to create a greater appetite for sweets? Whether or not the actual pathways of the brain are affected by exposure to the superexciting video games and entertainment available on screens, the mere difference between the slow pleasures of real life and the high-speed thrills of digital screens could cause an urge or taste for more excitement, a kind of *sensory-arousal addiction*.

As Dr. Richard A. Friedman, a professor of clinical psychiatry at Weill Cornell Medical College, wrote in the *New York Times* in 2014:

> [A]nother social factor that, in part, may be driving the "epidemic" of A.D.H.D. has gone unnoticed: the increasingly stark contrast between the regimented and demanding school environment and the highly stimulating digital world, where young people spend their time outside school. Digital life, with its vivid gaming and exciting social media, is a world of immediate gratification where practically any desire or fantasy can be realized in the blink of an eye. By comparison, school would seem even duller to a novelty-seeking kid living in the early 21st century than in previous decades, and the comparatively boring school environment might accentuate students' inattentive behavior, making their teachers more likely to see it and driving up the number of diagnoses.

So the real-world classroom has apparently become a boring environment compared with the stimulation and pacifying effects of popular devices. But the two are not mutually exclusive. There is a role for technology in the classroom. And the way to understand that role is to look at the role of cyber in the whole life of a child.

The Real Real-Life Experiment

On the jostling train to Galway, while the baby fed and the mother's eyes continued to be locked on her phone screen, I thought about all the complex events and conditions that are required for a human being to develop in the best possible way—the milk and nutrients needed, the sunshine, the balance of rest and stimulation and movement and sleep, and the hours of real tactile human contact with a patient caregiver. Raising a child is a daunting job, and it's no wonder so many parents surrender some of their duties to technology. So many other aspects of their lives are helped by it, made easier and better.

A baby's needs are not high-tech, though. And in many ways, technology has been proven to be less than beneficial for their healthy development. So far, no electronic device or app can replace cuddling, talking, laughing, playing a silly game, holding hands, or reading a book with your child. I have no doubt that someday tech developers and designers will create apps that can truly enhance learning for infants and toddlers, and then the educational value of screens will change. Until then, what we may need most is an app that reminds parents that they need to ditch their own screens at home and spend real face time with their kids.

Rather than developing swiping dexterity, which requires only a finger, babies need to do things with their entire bodies. They need to crawl, twist, and turn. They need to be bored! And they need to frown, grimace, and have a good, long, wallowing crying jag. They need to rise and fall. Achieve and stumble.

Embracing technology as an improvement over old-fashioned parenting will surely have an impact. Let's look at the chronological cyber effects of this over a child's lifetime. At birth, some babies are much less likely to experience face time and eye contact with their siblings, parents, and other adults. As they grow, they are commonly seeing adults around them reacting to their devices—and compulsively checking their mobile screens. Next, these children are being plopped in front of televisions and computer screens in their baby rockers, baby saucers, playpens, and Pack 'n Plays, or while suspended from ceilings on bungee cords. Anywhere but in their parents' arms.

Sherry Turkle describes a parent at a playground pushing a swing with one hand and using a mobile phone with the other. She has interviewed children who tell her that their parents read them a Harry Potter book in one hand while glancing at texts and email on a BlackBerry. And here's a telling story about school-age children—the subject of the next chapter—that Turkle gleaned from her fieldwork.

> Children describe that moment at school pick up—they will never tell you that they care—but will describe that moment that they come out of school, looking for that moment of eye contact, and instead . . . that parent is looking at the iPhone, at the smartphone, and is reading mail. So from the moment this generation of children met technology, it was a competition. And now they've grown up, and today's teenagers . . . now have their turn to live in a culture of distraction.

What memories are we building in our children?
What examples are we setting?

If we are cautious about what babies are exposed to, we won't be sorry down the road. There is no harm in waiting to find out.

And when it's time to introduce them to cyberspace, we should be looking for chances to share that environment with them rather than leaving them to navigate it alone. Put your four-year-olds on your laps and look at the screen together. If you are giving them a game to play, play it with them—and ask your children about it afterward, to see what they are learning. Clinicians should strongly emphasize the benefits of parents and children using interactive media together to enhance its educational value.

Monitoring is especially important once a decision has been reached to begin exposing an older child to screens. Whether this is television, computer games, a tablet app, or any other media, it is crucial for parents to question the appropriateness of the content. If your child is easily stimulated and often agitated, do not expose him or her to a highly stimulating show or app. All violence on digital media should be avoided, and when it is encountered, children should be helped to understand it. There are shows proven to calm and educate—or, at the very least, not overstimulate and cause insomnia, nightmares, or hyperactivity. I suggest the use of resources such as PBS Kids (pbskids.org), Sesame Workshop (sesameworkshop.org), and Common Sense Media (commonsensemedia.org) to guide media choices.

There are ethical questions to ask as well. Have tech companies and marketers of "educational" devices, apps, and games followed research-based child-development learning principles? Even if they have, many of these studies haven't been done with digital products and factored in developmental processes over time. I find it commendable that tech mogul Jeff Bezos has helped to establish a foundation that studies early child education, and recommends that more real books need to be purchased. A cynic might say that since he owns Amazon, and sells all those wonderful picture books that kids and parents love reading together, it's certainly in his interest to step out and say something. But the wisdom of this advice is irrefutable. When you are reading a book to your child, more than just an illustrated story is being expressed. It is about togetherness, affection, the creation of memories, bonding—and a thousand other small but profound things that occur during a moment of physical closeness. It is about nurturing and love.

And questions must be asked of the media. Studies show that despite the AAP recommendations and advice from experts, including an awareness campaign about early screen use in children that was launched by First Lady Michelle Obama, many parents are still in the dark. Most children are planted in front of a screen of some kind for four to five times the recommended amount of time. The latest surveys show that two-thirds of American families allow children under two to watch television and use tablets and mobile phone apps. According to the Kaiser Family Foundation, more than half of all children under two in American households watch between one and two hours of television or shows on computer screens per day.

This amount of time is a big part of that child's waking life, and using it in this way is likely to have adverse effects on their health and development. What can we do? There are a number of good resources online, including a Google+ community group that I'm involved with, Beyond the Screens, which gives news, articles, commentary, and up-to-date advice and recommendations by experts for parents.

In France, the government has led the way for the world, and in 2008 it banned programming aimed at infants in order to protect them from the negative effects of television. In fact, a warning now appears on foreign baby programs screened on French cable channels that (translated into English) reads, *Watching television can slow the development of children under three, even when it involves channels aimed specifically at them.* French schools offer awareness campaigns that discuss proper guidelines and suggested optimum ages for interacting with technology.

Taiwan has taken a giant and commendable step forward by recently introducing legislation to ban electronic devices for children under two and limit their use for those under eighteen—and hold parents responsible for any ill effects due to technological overexposure. Canada and Australia have similar recommendations and guidelines, but no bans.

This brings me to the United States, the leader in the development of digital technologies and products. It is a country where alcohol and tobacco consumption are monitored and limited, where asphyxiation warnings are written on plastic bags, and where warnings of choking hazards abound. It's good news that the AAP has announced that it is considering a revision of its recommendations on screen time for infants and young children, but so far, at the time of the publication of this book, no new recommendations have been made. In the meantime, babies turn into toddlers, who are quickly becoming young children. . . . Where are the warning labels on the tablet apps?

While we wait for more guidance and information or proper regulation, here are some ideas for parents:

- Don't use a digital babysitter or, in the future, a robot nanny. Babies and toddlers need a real caregiver, not a screen companion, to cuddle and talk with. There is no substitute for a real human being.
- Because your baby's little brain is growing quickly and develops through sensory stimulation, consider all the senses—touch, smell, sight, sound. A baby's early interactions and experiences are encoded in the brain and will have lasting effects.
- Wait until your baby is two or three years old before they get screen time. And make a conscious decision about the screen rules for them, taking into account that screens could be impacting how your child is being raised.
- Monitor your own screen time. Whether or not your children are watching, be aware of how much your television is on at home—and if the computer screen is always glowing and beckoning. Be aware of how often you check your mobile phone in front of your baby or toddler.
- Understand that babies are naturally empathetic and can be very sensitive to emotionally painful, troubling, or violent content. Studies show that children have a different perception of reality and fantasy than adults do. Repetitive viewings of frightening or violent content will increase retention, meaning they will form lasting unpleasant memories.
- Don't be fooled by marketing claims. Science shows us that tablet apps may not be as educational as claimed and that screen

time can, in fact, cause developmental delays and may even cause attention issues and language delays in babies who view more than two hours of media per day.

- Put pressure on toy developers to support their claims with better scientific evidence and new studies that investigate cyber effects.

And finally, how about doing a cyber-sweep of your baby's domain? Check the nursery, bedroom, and family room where they play—and banish anything with a digital chip or a blinking light. Resurrect the building blocks, the wooden trains and dolls. Go old-school! And when you play with your baby or toddler, get down on all fours (both of you), look into each other's eyes, and enjoy some unstructured quality time. Bond. Your baby will benefit enormously from this. And, believe me, so will you.

CHAPTER 4

Frankenstein and the Little Girl

Childhood as a concept has evolved over time. When did society decide childhood was precious—or that children had natural rights? In the seventeenth century, the English philosopher John Locke made a case for the sanctity of childhood by arguing that babies were born with a *tabula rasa,* or blank slate, and that it was the duty of parents to train and instill correct values and notions in them. The idea of the "vulnerable" child gathered more momentum during the Enlightenment, one hundred years later, when French philosopher Jean-Jacques Rousseau built on Locke's concept with added depth.

Childhood is an age of innocence, Rousseau argued, an important sanctuary in the life of all human beings before the harsh realities and perils of adulthood are encountered. He believed this time of innocence should be protected at all costs.

Not long ago, in terms of human history, most children left school once they were nine or ten years old. Boys were sent to work in coal mines and factories, or to apprentice in a trade like printing or blacksmithing. In 1818, nearly half the workers employed by cotton mills in the United States had started their jobs when they were under ten. To combat this practice, a succession of early child labor laws were introduced in America and Europe in the second half of the nineteenth century, but even so, these just restricted the length of a workday for a child to sixteen hours. At the time, one-third of the labor force of America consisted of boys under fifteen years old.

How a child is treated—how tenderly, how carefully—has always greatly depended on his or her parents. What if parents, whether due to lack of resources, lack of information, lack of interest, or in some cases just pure neglect, don't protect their children from harm?

As a moral and social issue, this problem was addressed during industrialization when enormous numbers of poor children, as young as three to five years old, were sent by their parents to work in dangerous environments—crawling up chimneys and into mines—for brutally long hours. This gave rise to the first child labor movement, which gathered strength and a wide array of crusaders in the nineteenth century, from English novelist and social critic Charles Dickens, who wrote popular books such as *Oliver Twist*—depicting the life he had known himself as a twelve-year-old boy forced to work in a shoe-blacking factory to support his family—to German philosopher Karl Marx, who criticized the British and U.S. economies for surviving on the "capitalised blood of children."

Even with a movement afoot and outspoken advocates the likes of Dickens and Marx, social change happened slowly. It wasn't until more than sixty years after the publication of *Oliver Twist* that the National Child Labor Committee was established to begin advocating on behalf of children in the United States—followed eventually by the Child Labor Coalition, Campaign for Labor Rights, and the United Nations International Children's Emergency Fund (UNICEF).

For the most part, the fallout from the first two industrial revolutions has been dealt with. But we have now entered a third industrial revolution—the digital revolution—and we are charting new territory regarding the care and treatment of children, and once again we are struggling with how best to protect "the age of innocence."

Will it take another hundred years to sort out? I hope not.

This chapter examines the online lives of children four to twelve years old. They are, in terms of harm and risk, the age group that is most vulnerable on the Internet. They are naturally curious and want to explore. They are old enough to be competent with technology, in some cases extremely so. But they aren't old enough to be wary of the risks online—and, more important, they don't yet understand the consequences of their behavior there.

Cyberspace has some very dangerous neighborhoods, yet it is by and large unmonitored. As a playground for adults, it's been compared to the Wild West. As a playground for children, it is entirely unsuitable. Yet in the frantic rush toward new technology, children have been forgotten.

"You wouldn't take your children and leave them alone in the middle of New York City," John Suler has said, "and that's effectively what you're doing when you allow them to go into cyberspace alone."

According to the journal *Pediatrics,* an overwhelming majority (84 percent) of U.S. children and teenagers have access to the Internet, on either a home computer, a tablet, or another mobile device. And as a marketing study shows, more than half of U.S. children who are eight to twelve have a cellphone. In fact, a 2015 consumer report shows that most American children get their first cellphone when they are six years old.

Six years old. This shocks me. This is before what in psychology we call the *age of reason,* when a child enters a new state of logic and begins to understand the surrounding world—learning the difference between right and wrong, good and bad, justice and injustice.

Now, with a cellphone in hand, these children are being catapulted into cyberspace before they are psychologically capable of making sense of it. And *we* can't even make sense of it yet. What this means in terms of their physical, cognitive, and emotional development remains largely unknown. It's still something we can only speculate about.

We do know, though, that technology has changed childhood in innumerable ways. Children are growing up with a digital foundation and framework for their lives. You might even call this new environment their true home. It is a place many of them have been visiting since they were very young and where, according to some studies, many spend an inordinate amount of their waking day. Cyberspace is where they are learning to read, doing their schoolwork, dressing up avatars, watching cartoons, and meeting friends both fictional and real.

Certainly there are vast and powerful educational aspects to this huge shift. This generation will build the technologies of tomorrow, so the positive aspects of networking and honing IT skills are clear. One could argue that the more time these children spend online, the better. In fact, it's been shown that those who spend a lot of time at their computer keyboards have more refined small-motor skills and enhanced hand-eye coordination. And a study conducted in 2009 by Beverly Plester and Clare Wood, both behavioral scientists in the U.K., demonstrated a positive relation between texting and literacy. In other words, reading and writing skills can be improved by texting, rather than ruined, as was feared by some. My own thinking is that a text is a form of coded language. It is encrypted. And receiving and sending such a communication involves both decryption and encryption processes. Since cryptography demands the highest level of cognitive function—think of Alan Turing's cracking of the Enigma code in World War II—I think if your children are texting a lot and they are really good at it, chances are they are pretty smart.

On the other hand, some developmental downsides of persistent and pervasive use of technology are already apparent, as noted in the previous chapter. Jo Heywood, a headmistress of a private primary school in Britain, has been outspoken about her observation—shared by other educators—that children are starting kindergarten at five and six years old with the communication skills of two- and three-year-olds, presumably because their parents or caregivers have been "pacifying" them with iPads rather than talking to them. This is seen in children from all backgrounds—disadvantaged and advantaged.

"Some [children] cannot even talk, let alone be in a position to learn to read via a phonics system, many teachers have reported," says Heywood. "This puts even more pressure on a primary system which is geared towards helping children read and write in the early years."

A report from a professional on the front lines is a very useful early warning system. And we should pay attention, even if parents tend to be enormously sensitive to criticism about their parenting habits, something I know firsthand. When I argued in early 2015 that mobile phones had become "virtual" babysitters (or "childminders," as we say on the other side of the Atlantic), it struck a nerve—and landed me on the front page of the *Daily Mail.*

If you find yourself questioning the dangers of early digital activity and insist on hard evidence backed by science, then you'll have to wait for another ten or twenty years, when comprehensive studies—the kind that track an individual's development over time—are completed.

Do parents really want to wait that long to find out?

Total abstinence from cyberspace is not the answer either. Years from now, it will be common cyberexperiences that this generation will bond over. Their fondest childhood memories could be made in a virtual playroom, completing a certain Mario game or their hours of exploring Club Penguin. But what other things did they experience in cyberspace? What other memories were made there, perhaps not-so-cheerful ones?

Social media scholar danah boyd (she uses all lowercase letters) has made an argument for not overprotecting children on the Internet, saying that they benefit from exposure the way they benefit from learning to roller-skate on the sidewalk despite the bruises and skinned knees. This thinking aligns with the child-rearing philosophy that teaching "resilience" of any kind is useful. I'll address the subject of resilience later in this chapter, but for now I'll say that bumps and bruises heal, but psychological and emotional wounds of childhood are more complicated and heal less easily. Do we know enough about cyberspace to declare it a safe place for exploring kids?

So what's the solution?

This is what caring parents often ask me while looking a bit afraid to hear my response. I appreciate how difficult this subject is for today's parents, who weren't raised with mobile phones and tablets and Wii remotes in their hands at the age of four or five. They can only make assumptions based on common sense, assisted occasionally by media reports or suggestions made by a pediatrician, schoolteacher, or therapist. But when it comes to technology, the experts are often frantically playing catch-up themselves. Believe me, they don't know the answers either.

Behavioral scientists aren't supposed to make assumptions. We rely on studies. So let's start there.

Nobody Knows If You're a Puppy

Human psychology is the study of human behavior. That means psychologists look at how people behave and what motivates them—and how people spend their time. So as we begin a discussion of four- to twelve-year-olds, it's important to first see how they are spending time in the cyber environment and what they are actually doing there.

Two recent studies, one of U.S. kids and another of European, tell us a lot about this. Between 2011 and 2014, a group called EU Kids Online conducted comprehensive studies, looking at children in twenty-two European countries and across many cultures. A strong majority of children used the Internet to visit social-networking sites like Facebook and to watch video clips on sites like YouTube. About half used the Internet for instant messaging and to do schoolwork. About one-third used it for Internet gaming, slightly less to download movies or music, and less again to read the news.

A similarly comprehensive study was done in the United States in 2014 by four academic researchers from the fields of education and psychology. A national sample of 442 children between the ages of eight and twelve, or what is called "middle childhood," were asked how they spent their time online. Younger children (eight to ten years) spent an average of forty-six minutes per day on a computer, compared with older ones (eleven to twelve years), who spent one hour and forty-six minutes per day on a computer.

When asked what kinds of sites they visited, YouTube dominated significantly, followed by Facebook, and game and virtual-world play sites—Disney, Club Penguin, Webkinz, Nick, Pogo, Poptropica, PBS Kids—all designed for this age group—and Google. Children with mobile phones (14 percent of eight- to twelve-year-olds in the study) played a lot of *Angry Birds,* a game that started as a phone app and is still primarily accessed that way.

Angry Birds, Club Penguin . . . that sounds fine, doesn't it?

But wait a second. What about Facebook? Don't you have to be thirteen years old to activate an account? Yes, but guess what? One quarter of the children in the U.S. study reported using Facebook even though it is a social network meant for teenagers and adults. These are the hidden users of social networks, the ones who aren't supposed to be there—but are. I think of them as "the Invisibles." It wasn't just eleven- to twelve-year-olds who were going there: 34 percent of the Facebook users in the study were eight- to ten-year-olds. In the EU study, one-quarter of the nine- to ten-year-olds and one-half of the eleven- to twelve-year-olds were using the site as well: Four out of ten gave a false age.

Twenty million minors use Facebook, according to *Consumer Reports;* 7.5 million of these are under thirteen. (But this 2011 study is already out-of-date. I wonder what the figures are now.) These underage users access the site by creating a fake profile, often with the awareness and approval of their parents. The technology editor of the *Consumer Reports* survey was troubled by the fact that "a majority of parents of kids 10 and under seemed largely unconcerned by their children's use of the site." Instagram has similar issues. The vast majority of the site's reported 400 million users are a young demographic, between eighteen and twenty-nine years old, but studies report that it is the most-used photography site for twelve- to seventeen-year-olds.

Identity and age verification online are complex issues. One of the popular jokes about this comes from a *New Yorker* cartoon that ran in 1993. The cartoon shows a dog sitting in front of a computer, and underneath the drawing, it says: "On the Internet, nobody knows you are a dog." It would appear that nobody knows if you're a puppy either.

Setting the minimum age for Facebook and Instagram at thirteen years is a data-protection requirement by law in the United States, but this doesn't appear to be strictly enforced. Why? In terms of scale, Facebook has 1.65 billion active members (as of May 2016) who make one post a day on average, including the uploading of 300 million images. Could these companies monitor and police illegal use of the site? When asked, Simon Milner, a senior executive with Facebook, said that it would be "almost impossible."

Almost impossible. Interesting language. If you can't police your own rules, then you might consider rethinking, revising, or removing them. Or simply closing down. That goes for the other social-networking sites as well. Patricia Cartes, head of Global Safety Outreach, Public Policy at Twitter, admitted that the company does not know which users are under thirteen. "We haven't found a silver bullet in the online industry," she said, "to meaningfully verify age."

Silver bullet? Like the one used to kill a werewolf?

Facebook and other social networks have always claimed that it is difficult—or "almost impossible"—to identify a child, and therefore they can't actively implement and police their own rules. But let's think about this for a moment. When a kid opens up a Facebook account, the first thing he or she typically does is put up a profile photograph, and then "friend" a bunch of schoolmates who are usually the same age. They go on to post comments about school, classmates, and extracurricular activities. If you can't figure out that these kids are nine or ten, you aren't very smart. They are constantly providing photographic evidence of their age. Another piece of evidence that makes me suspect that these social-networking sites are not particularly interested in monitoring this problem: In 2016, Facebook awarded $10,000 to a ten-year-old boy from Finland, a coding ace who discovered a security flaw in Instagram. Won't this only encourage more underage use?

I would argue that, when it comes to minors, there is an urgent need to develop more effective ways of verifying the age of a new user on social networks. The real-world example would be a liquor store or a pub that's not allowed to sell alcohol to underage individuals. Would it be okay if the salesclerk or barman didn't believe it was necessary to ask for proof of legal age for drinking—or wanted the profits more than he wanted to obey the law?

The psychologists and educators behind the large U.S. study in 2014 concluded that the results were troubling, particularly in regard to the developmental repercussions of children's online habits. "Engaging in these online social interactions prior to necessary cognitive and emotional development that occurs throughout middle childhood could lead to negative encounters or poor decision-making. As a result, teachers and parents need to be aware of what children are doing online and to teach media literacy and safe online habits at younger ages than perhaps previously thought."

The Bystander Effect

Obviously quite a number of parents are simply looking the other way. Perhaps they are quietly relieved, even proud, to see that their children are making "friends," usually a sign of social thriving and happiness. I think they need reminding about how ramped up the cruelty can be online. If you think girls of middle school age have always been mean, you've not seen what they can do in the escalated environment of the Internet.

Let's remember the story of Sarah Lynn Butler, the vivacious and beautiful twelve-year-old girl who was voted queen of the upcoming fall festival at her school in Williford, Arkansas, in 2009. The seventh grader was "very happy" with the news, her mother told the media, and "always laughing and giggling and cutting up and playing around." According to her mother, Butler had "lots of friends."

The problem was, after Butler was crowned, she began receiving mean messages on her MySpace page. Rumors circulated on the social network that she was really a "slut," along with other nasty descriptions. When her mother saw Butler's MySpace page and asked her daughter to talk about it, she was promptly removed from the friends list, and therefore denied access to her daughter's page.

Not long afterward, when the family left to run errands one afternoon, Butler asked to stay home. A browsing history revealed that she had logged on to her MySpace page and apparently seen the last message posted there, which said she was "just a stupid little naive girl and nobody would miss her." When her family returned later that day, they found her dead. The twelve-year-old had hanged herself. Her suicide note said that she couldn't handle what others were saying about her.

The stories of self-harm, even suicide, are growing in number—and, of course, the subject of *cyberbullying* has become an international conversation. In a poll conducted in twenty-four countries, 12 percent of parents reported their child had experienced cyberbullying—which is defined as repeatedly critical remarks and teasing, often by a group. A U.S. survey by *Consumer Reports* found that 1 million children over the previous year had been "harassed, threatened, or subjected to other forms of cyberbullying" on Facebook.

What is the explanation for it?

In general, the younger you are, the number of friends you have on a social network increases. Let's look at how the numbers work on Facebook, in a 2014 study of American users. For those over sixty-five years old, the average number of friends is 102. For those between forty-five and fifty-four years old, the average is 220. For those twenty-five to thirty-five years old, the average is 360. For those eighteen to twenty-four, the average is 649. What does that mean for the under-thirteens, the social media Invisibles? The answer is, Who knows? There are no reliable numbers.

Let's for a second discuss the sheer social madness of that. As the work of Robin Dunbar, a psychologist and anthropologist at the University of Oxford, has argued, primates have large brains because they live in socially complex societies. In fact, the group size of an animal can be predicted by the size of its neocortex, especially the frontal lobe. Human beings, too, have large brains because we tend to live in large groups.

How large? Given the size of the average human brain, the number of social contacts or "casual friends" with whom an average individual can handle and maintain stable social relationships is around 150. (It is called *Dunbar's number.*) This number is consistent throughout human history—and is the size of the modern hunter-gatherer societies, the size of most military companies, most industrial divisions, most Christmas card lists (in Britain, anyway), and most wedding parties.

Anything much beyond Dunbar's number is too complicated to handle at optimal processing levels.

Now imagine the child who has a Facebook page and an Instagram account, who participates on Snapchat, WhatsApp, and Twitter. Throw into that mix all the mobile phone, email, and text contacts. A child who is active online, and interested in social media, could potentially have thousands of contacts.

We are not talking about an intimate group of friends.

We are talking about an army.

And who's in this army? These aren't friends in any real-world sense. They don't really know and care about you. They are online contacts—their identity and age and name potentially false. According to Dunbar, if children have grown up spending most of their social time online with thousands of these "friends," they may not get enough real-world experience in handling social groups of any size, but particularly on a large scale—rendering them even less able to cope with real-world crowds. In other words, spending more time on social media can render children less competent socially, not more.

In the real world, if five friends turn on you, it is bad enough. Now imagine your class of twenty turns on you, and then imagine the entire school of five hundred kids turns on you. It would be unbearable—and you would stay home "sick" and hide under the covers. But now, imagine one thousand of your social-network "friends" chanting and pointing at you. Not many eleven-year-olds have the social skills to deal with that. Me neither.

Even if comments do not qualify as cyberbullying, a child of this age can be hypersensitive to criticism and, like a teenager, tend to focus on the cutting remark rather than the compliment.

I have been involved in two cyberbullying prevention campaigns for the EU's Safer Internet Day. In each case I've tried to employ some creative thinking and social science theories to come up with solutions. Because, it seems to me, all the money, time, and prevention campaigns for cyberbullying haven't really curtailed the incidences of abuse, either because they aren't working or because the number of kids going online without proper guidance is growing so exponentially that no program can possibly keep up.

For the first campaign, I used the *bystander effect* in psychology to raise awareness of cyberbullying among schoolchildren. The bystander effect refers to a crime that took place in New York City in the 1960s, when a young woman, Kitty Genovese, was being stabbed to death and cried out on the street for help, but nobody came. After studying this disturbing case, psychologists learned that the greater number of people who witness a crime or emergency, the less likely any of them will feel responsible to respond. There's another term in psychology for this phenomenon. It is called *diffusion of responsibility*. The tenet is similar. When part of a large group, everybody thinks that someone else will act.

Think how this may work online. In the case of cyberbullying, hundreds of "friends" can witness bullying or harsh criticism online but don't step up and do anything. It's actually possible that *the more friends you have, the less likely somebody will intervene*. For this campaign, the motto that I created was: "Don't Be a Bystander—Stand Up and Do Something." In other words, don't wait for one or two of the other three hundred people to do something. It's up to users to create a better environment.

The second campaign was called "Be a Cyber Pal," a bullying prevention initiative for Safer Internet Day in 2014. Again, I employed a sound theoretical approach from psychology, the *theory of planned behavior* (TPB). This means that the more you mention something, the more you normalize it. So my initiative was to do an anti-cyberbullying campaign without mentioning the actual word *cyberbullying.* I can't help but think that the more we mention it, the more we normalize it, and therefore may increase the probability and expectation of kids being cyberbullied.

"Be a Cyber Pal," conceived as an antidote to cyberbullying, was about actively being a kind, considerate, supportive, and loyal friend. And it is cause for hope that it became the most downloaded poster of the campaign that year. I think the positive message—rather than repeating another scary cyberbullying story—gave teachers and families something that's easier to talk about.

More recently, hoping to apply solid science to other solutions, I have been working on a mathematical formula to predict the prevalence of antisocial behavior online—in hopes of designing an algorithm to identify incidences of bullying. How?

Locard's exchange principle is the basic premise of forensic science. As I mentioned in the prologue of this book, it dictates that every contact leaves a trace, and nowhere is this more true than online. Unlike the playground, where the mean words of a bully disappear instantly into the ether—unless there is an eyewitness—online it is just the opposite. Cyberbullying is nothing but evidence: a permanent digital record. So how did we get to the point where it became more problematic than real-world bullying? My answer is taken from *The Usual Suspects,* one of my favorite movies, in which Kevin Spacey delivers the immortal line "The greatest trick the devil ever pulled was convincing the world he didn't exist."

To me, the greatest trick social media and telecom companies ever pulled is trying to convince us that they can do nothing about cyberbullying.

In terms of digital forensics, it is a cybercrime with big fingerprints. Using an approach that I am calling the *math of cyberbullying,* both victims and perpetrators can be identified.

Many of the big-data "social analytics" outfits like Brandwatch, SocialBro, or Nielsen Social use algorithms to identify or estimate much more complicated things, like a Twitter user's age, sex, political leanings, and education level. How hard would it be to create an algorithm to identify antisocial behavior, bullying, or harassment online? My equation goes like this: *d* x *c* (*i* x *f*) = cyberbullying.

The math would be this simple:

I am bullying you = direction (d)
bitch, hate, die = content (c)
interval (i) and frequency (f) = escalation

I am actively working with a tech company in Palo Alto to apply the *Aiken algorithm* to online communication. To develop the c (content) database, I plan on launching a nationwide call for content. Every person who has ever received a hateful bullying message can forward it to our repository. In that way, victims of cyberbullying can become an empowering part of the solution to an ugly but eminently solvable big-data problem. We just need the collective will to address it.

The algorithm can be set to automatically detect escalation in a cyberbullying sequence, and a *digital outreach* can be sent to the victim: "You need to ask for help. You are being bullied." And simultaneously an alert can be sent to parents or guardians telling them something is wrong and encouraging them to talk to their child.

The beauty of the design is twofold—first, only artificial intelligence would be screening the transactions, which will be incredibly efficient for a big-data problem such as cyberbullying, and second, there would be no breach of privacy for the child. Parents wouldn't need to see the content, only be alerted when there appeared to be a problem. I know there could be an outcry about surveillance, but we are talking about minors and we are talking about an opt-in solution with parental consent.

Ultimately the algorithm could reflect jurisdictional law in the area of cyber-harassment against a minor and be designed to quantify and provide evidence of a crime. One day, it could involve sending digital deterrents to the cyberbully, which is a way to counter what cyberpsychologists call "minimization of status and authority online." We can show young people that there are consequences to their behavior in cyberspace.

It's a twenty-first-century solution to a twenty-first-century problem.

Nastiness online is becoming an accepted reality—and something most people have witnessed. The majority of adult social media users said they "have seen people being mean and cruel to others on social-network sites," according to a report from the Pew Research Center's Internet and American Life Project. The conditions of the cyber environment can make cruelty a competitive sport—and posts escalate from barbs to sadism very quickly. Envy drives some of this activity. Celebrities are often targets. It took Monica Lewinsky, one of the early social media victims, a decade to emerge from her experience of being shamed and humiliated. Zelda Williams, the twenty-five-year-old daughter of actor Robin Williams, gave up her Twitter account after she was exposed to unimaginably awful tweets following her father's death.

The same year, when American baseball legend Curt Schilling tweeted his paternal pride that his daughter Gabby had received a college acceptance letter, the celebratory mood devolved into ugliness when Twitter "trolls" engaged in sexually explicit posting about Gabby, who was seventeen years old. Schilling did what probably millions of other fathers can only dream about: He used his fame and popular blog to track down nine of the individuals who had generated the hateful and sexist comments and got them fired from their jobs or sports teams.

If young adults can be so devastated by online attacks—then what about children?

"Trolls" are malicious individuals who search online for unsuspecting people to deceive and trick. Sadistically teasing and taunting children and tweens is a sick sport for them. One common place where they meet up with kids as young as six years old is on Internet gaming sites, where groups use webcams and microphones to communicate while they meet one another online and play. They can be found on popular multiplayer online games like *Grand Theft Auto* (affectionately known as *GTA*), which they play in hopes of winning the trust of young unsuspecting players in order to trick them, usually making them freak out—while recording their conversations and posting them for kicks. This is damaging for children on so many levels, not to mention that it brings them into contact with these pathological strangers who are manipulating and preying on their innocence—for laughs.

The Elephant in the Cyber Room

But let's dig a little further into the comprehensive EU study. Children were asked if they had been bothered or upset by anything they'd seen online—or knew about things that bothered their friends.

Bothered was used to describe something that "made you feel uncomfortable, upset, or feel that you shouldn't have seen it." The children were asked to describe in their own words what bothered them.

Nearly ten thousand responses came in. They were diverse and wide-ranging, and changed considerably with the age of the child. Younger children were more concerned about content, such as something they'd seen that was meant for adults. Older children were more worried about conduct and contact, in other words, troubled by something they'd done or witnessed being done online. They worried about how they were supposed to act (conduct) and about people they might meet (contact) online.

The researchers compiled the responses and organized them into types of concerns. Girls tended to be more concerned about strangers they had met online, or might meet. Boys were more bothered by violence they'd seen. Both boys and girls described being bothered by things they'd seen on video-sharing websites like YouTube—violent or sexual images, as well as other inappropriate content. Both boys and girls described being bothered by real violence, as well as gory, cruel, and aggressive fictional violence that they'd seen in other places online—particularly violence against animals or other children.

Okay, let's stop there for a second. *Children watching violence against children?* Yes, you read that correctly. Almost anything can go up online, on any forum or site where video hosting is enabled. Some sites do monitor content, but there is often a latency period, or window in which unmonitored content can be seen by anyone before it is taken down.

It's bad enough to watch violence against adults or animals, but what is the impact of children seeing violence against other children? There is actually a real-life forensics case that involves this. A two-year-old boy in Britain, Jamie Bulger, was kidnapped in 1993 by two ten-year-old boys in a shopping center while Jamie's mother wasn't looking. The older boys tortured and murdered Jamie; his mutilated body was found a couple of miles away. Here's the piece that connects the act of witnessing violence with behavior: It was reported that the two boys had been actively looking for a young child to torture after they had watched horror videos, specifically the 1991 film *Child's Play 3*. It was claimed that the boys staged Jamie's death, even mimicking some of the brutality in the movie. While this awful and tragic case predated the Internet, we can learn from it today.

Let's return to the study of eight- to twelve-year-olds and what troubled them online. What bothered them most?

Pornography they'd seen.

Yes, that was the most frequently reported concern. As a reality check, in the same study, 5 percent of the nine- and ten-year-olds had seen sexual images online in the previous year. In other words, when they were eight and nine.

Pornography. *This is the elephant in the cyber room.* We aren't just talking about nudity. It's adult pornography and hardcore pornography that is "upsetting, intrusive, or inappropriate."

Did we need a study to inform us of this? Have we reached a time in history when we have to ask children what bothers them in order to find out that pornography isn't for eight- to twelve-year-olds? This is a terrible moral blind spot of our society.

As much as I trust empirical science and respect the care given to creating studies that accurately reflect reality—when it comes to the subject of the cyber effects of online pornography on the developing child, I am not sure that we have time for the careful studies that would provide definitive results and information. And even if we did have time for studies, it would be almost impossible to conduct them, due to the ethics of exposing children to harmful material for the sake of a study. In many ways, we are simply in the dark. And we may be there for the foreseeable future.

But does that mean we do nothing?

It was social psychologist Leon Festinger who created the theory of *cognitive dissonance*. It occurs when our thoughts or ideas about the world clash, and we feel tension and internal conflict, which is unpleasant for us. Seeking internal harmony, we make a choice—often not consciously—to focus on something other than the conflict, which could be irrational and maladaptive behavior. The elephant in the cyber room reflects what I think of as *societal cognitive dissonance,* the experience of knowing intuitively that something, like the Internet, is both good and bad for society—but choosing to ignore the bad effects in order not to feel conflicted. Technology causes this trade-off. We want to reduce tension and internal conflict, and to focus on all the nice things the Internet brings us—Wi-Fi, connectivity, convenience, fun gadgets that enhance status. We decide to ignore the problems and risks.

The bystander effect also applies. We are all witnessing this crime against innocence. Does the size of the crowd online cause each of us to look the other way?

The Age of Exploration

Imagine that a nine-year-old child is wandering around a 7-Eleven—or any regular convenience store of the sort that urban dwellers have come to know and rely on because, well, they are so convenient. In Japan, it's a FamilyMart. In Norway, it's a Narvesen. You can get cash, milk, energy bars, eggs, juice, a newspaper, and a cup of not-so-great coffee there.

Candy takes up an entire aisle. Another aisle has magazines.

When you were a child, you may have visited a convenience store with your friends or parents. You stood and looked at the candy aisle. You may have browsed some magazines—flipping through the pages of *Outside, PC World, Popular Mechanics, Newsweek, Seventeen,* or *Cosmopolitan* with its racy covers.

A nine-year-old child has some grown-up interests. They may still believe in Santa Claus and fairy tales, but they are also curious about the adult world. It terrifies them a little but also beckons.

Psychologists agree that exploration is a healthy and necessary part of development. Emotionally, a nine-year-old child is developing self-regulation and self-control, beginning to pull away from family and parents by developing closer bonds to peers at school. They are beginning to develop a "self-concept" and self-esteem, usually by evaluating self-worth in comparison with others.

Around the world there are different cultural mores about what is appropriate for a child's eyes, but by and large, wherever you go, adult content—words and images—is kept from children. This is often regulated by the government in some way, the result of laws that have been written, passed, amended, and honed for decades.

In the convenience store, the adult magazines—*Penthouse, Hustler, Playboy*—are kept under opaque cloth or colored plastic so you can't see their covers. They might be kept in a separate room—in the way that adult movies, before the era of Netflix and On Demand, were kept in a different room of a video store that children were not allowed to enter.

Prior to the Internet, it was quite possible that a nine-year-old child would not have been aware of the existence of adult magazines or where to obtain them—and even if they did, obtaining them wasn't easy. Now imagine that nine-year-old child sitting in front of a computer screen. Perhaps the child is clustered with a group of friends, but more likely alone.

Parents would be vigilant in a city convenience store with a child, and keeping an eye out. But at home they are not so vigilant. What is there to be afraid of? Yet within a few seconds of searching on the Internet, a child can come into contact with the adult world in a shocking variety of ways—which can leave them disturbed or, as the EU study says, "bothered." They might be afraid to say anything to their parents for fear of getting in trouble—or worse, having their devices confiscated.

This is not the same as an issue of *Playboy* hidden under a bed. Such a magazine might contain a total of sixty images or fewer, which would be (relatively speaking) pretty benign. Now in the time it once took to turn a page, there are thousands of websites with millions of images. There are challenges in quantifying content online, but in early 2016, according to the Internet Filter Review, which is continually updating its findings, there were 4.2 million pornography sites and 68 million daily pornographic search engine requests—or one quarter of all searches. The prevalence and accessibility of this content may be too great to avoid.

Some social scientists have even speculated that the rapid growth of the Internet, and its design, were driven in part by the great human attraction to and interest in pornography. Given the numbers of people who are into it—a 2014 *Cosmopolitan* survey of four thousand adult men found that 30 percent of them viewed it daily—we must consider it a relatively normal interest for adults.

What makes you think children aren't seeing it?

I wonder sometimes if the potential threat hasn't been made clear enough to parents: the impact of pornography on developing children. As Michael Seto described, it could be the largest unregulated social experiment of all time.

As a society, we are readily having conversations about the impact of Photoshopped images of women online and the impact on young girls, who tend to think of themselves and evaluate themselves in comparison with the appearance of those unreal and Photoshopped celebrities. Society discusses the impact of concussions on young boys who are playing football (though it took decades). Society discusses and amends the appropriate age to begin driving a car, drinking alcohol, voting in federal and state elections, joining the military, and getting married.

And yet society does not appear ready to have this conversation: What is the effect of explicit and upsetting images of adult sexual practices and pornography on developing children?

While many parents do describe themselves in surveys as concerned about pornography on the Internet, it is enormously troubling to me that not all of them do. For instance, when parents of young children in the U.K. were asked what worried them about raising children in the digital age and were given a list of things that could concern them—the list included children accessing pornography or sharing self-generated sexts—15 percent of them said, nope, they weren't concerned about that.

Incredibly, as many as 11 percent of all parents surveyed expressed concern that the technology skills of their under-five-year-olds surpassed their own. It goes without saying that if your tech skills aren't as sharp as those of your five-year-old child, how are you supposed to protect them online? Eight in ten parents feel they do not know enough to keep their three- to four-year-old children safe online, and the same goes for parents of five- to seven-year-olds and eight- to eleven-year-olds.

Kids are curious. And when they become curious about something online, they will find a way to see it and share it with other kids. Older children—older than twelve—may have the self-control and self-knowledge to resist the temptation to see something disturbing. But for a younger child, simply due to where they are developmentally, this is often not possible. Which brings us to the important subject of parental control.

Parental Loss of Control

I am often asked about parental controls: *Which ones are best?* and *Do they really work?* Shopping for this software can be confusing—like shopping for antivirus ware—and expensive, with monthly fees and updates and occasional bugs. There are plenty of offerings available, which are rated and reviewed online by consumers and parent groups.

My concern isn't so much for the capable parents who are good at researching and finding the best options; it's for the overburdened ones who don't have the time, money, patience, or ability to seek the best options, much less install them. Even after the software is installed, parental controls can give a false sense of security, lulling naive parents into a lack of vigilance. Many tech-savvy kids have lots of ploys they use to get around parental controls. It doesn't require much resourcefulness. If you want a real wake-up call, just go online and do a Google search for "ways to bypass parental controls," and you will find more than a million results. It is the ultimate expression of democratization of knowledge.

Each year, with the speed of technological changes and creation of new social media sites—I hate to start naming them because by the time you are reading this, there will be new ones—it becomes harder to adequately observe, study, and monitor the content that children are accessing online.

But it is the proliferation of handheld devices—a kind of privatizing of Internet use—that has had the most impact of all, making it unrealistic for parents to watch over their children's shoulders in order to keep them safe by limiting Internet access to a family room or public part of the house. Studies in the past have looked at the content of TV and devices as separate entities, but these have obviously merged now. Kids watch television programs, movies, and YouTube videos on their computers and mobile devices. At least there is parental control software for most mobile phone browsers and tablets now. For many years after the launch of the first iPod Touch, a small handheld device for music-listening and browsing of the Internet, no parental controls were available for it, even when the largest market for it was middle-school-age children. Eventually this was rectified, but millions were sold before any controls were created.

Apple cofounder Steve Jobs was known to keep a tight lid on the use of screens in his own household—and Silicon Valley's most tech-advanced parents seem to be the ones who are most strict about Internet access for their children. But even if you're a parent who feels confident that the filtering software for the household and mobile devices is working beautifully, what happens when your child goes to a friend's house?

Many parents simply give up. According to studies, they are overwhelmed by technology and feel they can't keep up with accelerating online advancements. Meanwhile, children take advantage of their parents' limited focus, lack of tech acumen, and lack of time. A large percentage of them "hide their participation in risky and sometimes illegal activities," as one study described.

Trying to be helpful to parents in the U.K., the government asked its independent regulator, Ofcom, to recommend proper monitoring of a child's online life after a study in 2014 found that one in twenty U.K. families with young children did no monitoring whatsoever.

A panel of experts came up with a four-point approach to protect children online:

1. Using technical mediation in the form of parental control software, content filters, PIN passwords, or safe search, which restricts searching to age-appropriate sites.
2. Talking regularly to your children about managing online risks.
3. Setting rules or restrictions around online access and use.
4. Supervising your children when they are online.

As it turned out, only one-third of families with five- to fifteen-year-old children at home provided all four forms of mediation. I can guess why. Because it's a full-time job!

How on earth can a parent be expected to do all that, while feeding, clothing, and caring for a family—and probably working forty hours a week? The figures vary from study to study, but in general, for homes with children who are a bit older, over twelve, the percentage of homes with no controls or monitoring rises to around one in ten.

As a forensics expert, I believe whistling in the dark is a gamble. In my field, there's a famous case, Situation 21, that speaks to this question. When organizers of the 1972 Olympic Games in Munich were trying to prepare for any possible security risks, they called upon West German psychologist Georg Sieber to imagine various worst-case scenarios to help plan for all contingencies.

Sieber came up with twenty-six possible scenarios. The one that he called Situation 21 described a scenario in which armed Palestinian terrorists invade the Israeli delegation's quarters, kill and take hostages, then demand the release of Palestinian prisoners from Israel. But this scenario and all the others that Sieber came up with seemed over-the-top and incredibly unlikely to the Olympic organizers. So dark and so negative! After all, the 1972 Olympics had been given the official motto *Die Heiteren Spiele,* which translates as "The Happy Games." To even consider Sieber's awful imaginings could ruin the fun, carefree spirit that the organizers hoped for.

Eerily, Situation 21 came true. During the Olympics, eleven Israeli athletes and one German guard were murdered. So much for "The

Happy Games." The phrase became a grim reminder that merely hoping for the best will not prevent the worst.

The reality: When a risk is unlikely—and unpleasant to consider—there is often a strong desire to overlook it. The explanation is that human beings must continue living, day in, day out, despite the risks we all face, whether it's a car crash, a theft, an act of terrorism, or a visit to your bedroom by a random serial killer with a hatchet. But with the help of laws, government regulations, and technology, we face fewer risks each year. Our cars have alarms, air bags, and better seat belts. The number of airplane crashes varies from year to year, but lessens overall. Same goes for the murder rate in most developed countries. The ability to track serial sex offenders using DNA tests and advanced technology makes them less of a threat too.

From a forensic psychology perspective, I can't help but sympathize with Sieber. The desire for the freedom in cyberspace, and the reluctance to consider its true downside, causes some parents to ignore the risks of what could happen to their children online, just as the Munich organizers in their state of cognitive dissonance refused to increase security or to consider Situation 21 because it would go against their desired concept of a relaxed, fun Olympics. Does the actual degree of risk mean that you shouldn't prepare for it—or discuss these difficult things with children?

My job, as I see it, is to be fully armed with real insights and information, both open-eyed and imaginative, about potential risks so I can be prepared for the worst-case scenario. As we say in risk assessment, "Start at the apocalypse and work back."

The variety of unsupervised and age-inappropriate content to explore online is almost limitless. And the number of children exposed to it grows every hour. This is a situation 2.0 that we cannot ignore.

Sweetie + Webcam Sex Tourism

Sweetie was a playful ten-year-old Filipino girl with large brown eyes and shiny chin-length hair. "Hello, my name is Sweetie," she said in singsong accented English as she looked into her computer webcam and greeted new friends.

Online, Sweetie was willing to have webcam sex for a price. Webcam "sex tourism" is a growing business, partially due to the hypervigilance about sexual offenders in the real world. The result: Men in rich countries pay thousands of children in poor countries to sit in front of a computer webcam and perform sex acts, or watch the men perform them.

In the first two and a half months of engaging with strangers in cyberspace, Sweetie attracted one thousand predators from seventy-one different countries. But before you get terribly depressed about this wretched story, let me pass along the good news: Sweetie wasn't real. She was an interactive 3-D computer model that looked and moved and sounded exactly like a real ten-year-old Filipino girl. She was the creation of Lemz advertising agency for the international children's rights network Terre des Hommes (Netherlands). Sweetie was designed to capture the identities of child predators and pass them along to law-enforcement agencies. At any time of day, according to Terre des Hommes, roughly 750,000 men worldwide are looking for online sex with children.

"The moment [Sweetie] got online, we were swamped, like an avalanche," said Hans Guyt, the special projects director of Terre des Hommes at the time.

The government of the Philippines estimates that there are as many as one hundred thousand children in the country who are forced into the webcam sex trade. Legally, the creation of an avatar like Sweetie to lure predators can be considered a form of entrapment in some jurisdictions, so this sort of practice does trouble me. But many countries, like the Philippines, in an effort to fight online abuse and the escalation of webcam sex trafficking, are changing their laws.

Sexual deviants who are drawn to children are broadly described in two ways: content or Internet offenders and contact offenders. One type is looking for content, or images of children, while the other is seeking physical contact with a child. The Internet facilitates both. How?

In the old days, a contact offender had to fly to the Philippines or Thailand to engage in sex with a child. Now they find their children online—in what is called *technologically facilitated abuse*—and by visiting forums and sites where pedophiles share information and a spirit of camaraderie.

"Cyberspace allows people with truly pathological traits to find similar comrades who reinforce their problem," John Suler wrote in *Psychology of the Digital Age: Humans Become Electric.* As discussed in chapter 1, the mathematics of online behavior comes into play. Due to the effects of online syndication and escalation, the availability of like-minded collaborators in a new environment, which socializes and normalizes the deviant behavior, compounded by the effects of online disinhibition and anonymity, it is likely that an individual with a predisposition for pedophilia may find that behavior harder than ever to control. Additionally, people with criminal intent are able to access victims more easily.

Once upon a time, a content abuser who collected indecent images of children had to keep a secret stash in a box hidden in the attic. Now the amount of material available to this population is rapidly growing, thanks to the ease of taking digital images with mobile phones and computers and then sending or posting them online. While *child pornography* is the legal term used, children's advocates prefer *child abuse material* (CAM), which removes any implication of consent of the child or a benign relationship with adult pornographic practices. Ten years ago, the National Center for Missing and Exploited Children (NCMEC) received thousands of images. In 2012 alone, the center received nearly 20 million images. As of 2016, the NCMEC archives had grown to a total of 139 million photographs in its Child Victim Identification Program. When a sample of images was assessed, 70 percent were classified as child pornography, 16 percent were described as online enticement, and 14 percent were classified as self-production. In other words, they were selfies—or sexts.

Interestingly, the profile of the average offender has changed in recent years. I suspect this has to do with technology as well. In the past decade there's been an emergence of younger online offenders. Some reports indicate that an increasing number of them are under the age of eighteen. What would cause such a phenomenon? Does the access to pornography online have anything to do with it, or more simply, the impact of an adult online environment on children?

Back in 2010, when I was researching online child abuse material and cowriting a paper with INTERPOL assistant director Michael "Mick" Moran, I came across interesting information in the studies. An inappropriate sexualized event in early childhood has been reported by some contact and online offenders. This alone makes me seriously concerned about children looking at adult material online, because I would argue that the viewing of inappropriate content itself qualifies as an "early inappropriate sexualized event." If so, the potential consequences could be tragic. As a society, we need to stop and really think about this.

Now, after the darkness of this difficult subject, I think it's time for some good news: Sweetie isn't the only clever development in recent years to fight pedophilia. Increasingly, the solution to technology-facilitated deviancy lies with technology itself.

"Every Image Is a Crime Scene"

For three long years, INTERPOL had been trying to catch a particular pedophile—a man who had posted more than two hundred photographs online that showed him sexually abusing young boys. All told, there were a dozen different boys in the photographs, their ages ranging from early teens to—and this is difficult stuff—as young as six years old. Judging from the features of the boys and a content analysis of the images, the location of these crimes appeared to be Southeast Asia.

Posting was a daring move for the pedophile, except that, in each of the shocking photographs, the face of the man had been digitally swirled, rendering his identity a total mystery.

Mick Moran was a detective at INTERPOL and working in the Crimes Against Children unit at the time of the investigation in 2007. Mick is a big, gregarious bloke, a former Irish Guard who shoots from the hip and is unrelenting in his quest to catch predators of children. He and I met around that time, and he invited me to become involved in research with the INTERPOL Specialists Group. He has a memorable saying about indecent images of a child or minor: "Every image is a crime scene."

You know how forensics experts comb over a crime scene to find evidence? Moran's team does brilliant analysis of child-abuse images to crack his cases. So when he was put in charge of the INTERPOL investigation of the pedophile with the "swirled" photographs online, Moran sent the images to a German lab, where cutting-edge technology was used to unscramble the features of the offender's face.

Acting quickly, an unprecedented global manhunt was launched. An INTERPOL worldwide appeal received nearly 350 tips from the public within days. Five sources on three continents—including confirmation from a member of the predator's own family—identified the man as Christopher Paul Neil, a substitute schoolteacher from British Columbia and a former chaplain and counselor with the Royal Canadian Air Cadets in Nova Scotia and Saskatchewan. Neil had left his studies for the priesthood (I know what you're thinking) to move to Asia, where he taught English for five years.

Neil was working in South Korea but fled to Bangkok when the manhunt began. "I know the Thai authorities," Moran told the media, "[and] all the countries in this region—Cambodia, Vietnam, Thailand—I know they have all been alerted." Border controls were in place and a collaborative effort to find Neil resulted in his arrest by Thai police.

INTERPOL announced in a statement that the operation would "serve as a warning to pedophiles that the power of the Internet, and public revulsion, would leave no hiding place." In the end, the very technology that was publicizing Neil's crimes brought him down.

"The irony that we used the Internet to publicize our message is not lost," Moran said.

It's a fact that technology has made it easier for pedophiles and other sexual deviants, but what can give us hope is the commitment and resourcefulness of individuals in law enforcement worldwide, like Mick and so many others I've met in the course of my work, dedicated to fighting the escalating problem of child abuse materials online. Together, with innovative new technologies and the determination of inspirational people like Sharon Cooper, a forensic pediatrician at NCMEC, and the creative vigilantes at Terre des Hommes and other NGOs and nonprofits devoted to saving as many children as possible, I feel confident that more pedophiles and other predators who market and trade child abuse material will be stopped. Outwitting those who prey on children is something we can cheer about in this dark world.

Although this vigilante stuff can definitely go too far, as in the instance of the unfortunate pediatrician in South Wales whose house was sprayed with graffiti by vandals.

Police confirmed that the attack was prompted by confusion over the words *pedophile* and *pediatrician.*

So, vigilantes, please watch your spelling.

Frankenstein + The Little Girl

All the gathered knowledge of human civilization is available by using Google, Bing, Yahoo, and the other search engines, and very soon all the books ever published throughout recorded history. The researching and self-teaching potential is awesome.

But not for a young child. Not even for children who are eight to twelve, even though a U.S. study reports that a large number of them are regularly conducting searches on Google.

The essential problem lies with the search algorithms. They are designed to speedily deliver listings of the most frequently searched phrases. Extreme content and scary scenarios, which always draw the most adult eyeballs, can be presented first—regardless of the searcher's age—due to the popularity of the sensational information.

So when a ten-year-old girl sits down in front of a computer and enters a word into the beguilingly empty search box, then presses "Return," that's when she is suddenly interfacing with machine intelligence. Almost anything becomes possible at this point—either by an accidental landing on an inappropriate site or by the profound power of a child's curiosity. Some of the listings could be truly irresistible to a young child. In school and on the playground, as well as online, word of mouth is responsible for half of the traffic to various sites, including problematic ones. You don't have to be a psychologist or expert on child development to know that children in this age group are frequently most interested in sharing the names of sites with inappropriate content—and daring one another to visit.

The technology that's facilitating the ten-year-old girl's search, as intelligent as it is, has no awareness that it is providing information to a ten-year-old. As a society, we haven't focused enough on the ethics of connecting machine or artificial intelligence with a child. When a man-built machine harms a human being, who is responsible? Is it industry, the designer of the technology, or the owner of the machine? (The first recorded death by robot was in 1979, when an assembly-line worker for Ford Motor Company was killed instantly when a robot's arm slammed him.)

It reminds me of that scene in the classic 1931 film *Frankenstein,* when the monster makes friends with a little girl who is picking flowers by a lake. The whole thing seems very sweet until he throws her into the water and she drowns.

Mary Shelley wrote her book *Frankenstein* amid the confusion and fears of industrialization—and speculated about what science might bring in the future. It is a perfect example of the power of fiction to present and illustrate a complex historical shift more effectively than dozens of nonfiction accounts. Would Shelley ever have imagined how the algorithms of search would bring machine intelligence into every-day life?

And where does this digital Frankenstein monster, online search, take the girl? While participating in an advisory group in Ireland to provide guidance about Internet governance, I was asked to study the offerings of legal but age-inappropriate content that could be accessed easily online by children. Believe me, it includes a great deal more than pornography and extreme violence.

It is actually very easy to find websites that promote drug-taking and alcohol abuse in the form of competitive drinking games. One troubling social media trend, *neknominate,* apparently started in Australia and swept the rest of the social media world in 2014. Other sites offer aggressive behavior in the form of extreme hostility: racism, hate speech, anti-LGBT (lesbian, gay, bisexual, and transgender) attitudes, or expression of extremist political views and attempts at radicalization. And let's not forget the pro-anorexia and pro-bulimia sites (so popular that there's even slang abbreviations for these: "pro-ana" and "pro-mia") that promote life-threatening eating disorders as a lifestyle option. Worst of all, there are many self-harm (cutting) and suicide websites that target young people and provide a forum where they can meet other kids who engage in self-harm and therefore reinforce and normalize the behavior.

Imagine our ten-year-old girl discovering one of those sites. Perhaps she's been hearing a lot about obesity at school and on TV. U.S. First Lady Michelle Obama has been a vocal activist on the subject, and obesity has been a trending topic for the past few years. Developmentally, a ten-year-old girl is beginning to think about herself in a new way—in relation to the other girls at school, and in comparison with the girls she sees online, in magazines, and on TV. She knows she's a little chubby. Her aunts and grandmother are always pinching her apple-cheeks and telling her so. Is she obese? She may worry about that, so she enters the word "thin" and finds links to pro-anorexia and pro-bulimia sites (sometimes described as "thinspiration") within seconds.

We also know from research done previously regarding television about the importance of children watching age-appropriate content. Young children can react very positively to age-appropriate shows and lessons—but will react just as negatively if the content is inappropriate for them. What kind of negative reaction? As I've said, studying the impact of pornography or violence on children is problematic to conduct due to the ethics of experimenting this way with the minds of minors. But when children are surveyed about their experiences online, we can draw some conclusions from the results. A child helpline in the U.K. survey in 2015 reported that one in ten children who were twelve to thirteen years old were "worried they might be addicted to porn." One in five of those who were twelve to seventeen said they had seen pornographic images that had shocked or upset them, and disturbingly 12 percent said they had taken part in, or had made, a sexually explicit video.

More recently, an editorial by Alice Thomson in the *Times* of London highlights the ubiquity of hardcore pornography and what this could mean for children. One site, Porn Hub, had videos watched 87 billion times in 2015—the equivalent of twelve videos per person globally. As Thomson points out, police are struggling against a tide of offenses fueled by the easy availability of pornography, always just a few clicks away, even from children. In 2016, the governor of Utah, Gary Herbert, signed a resolution declaring a "public health crisis" due to its prevalence and easy access online. This is a societal crisis.

Content moderators of the Internet—adult individuals who are hired to remove beheadings, tortures, rapes, and child abuse images from search and social media sites—do report serious emotional fallout from what they've seen while doing their work. The ill effects on them—insomnia, anxiety, depression, recurrent nightmares—are very similar to those of people suffering from post-traumatic stress disorder, something we dramatized in a second-season episode of *CSI: Cyber.* The episode was called "5 Deadly Sins," referring to the five types of content that social media companies pay to have removed (hate speech, porn, violence, drugs, trolling). I find it very disturbing that moderators are being used as human filters for extreme content. This is certainly an area crying out for a technology solution.

Moderators and deadly sins aside, we all know that the Internet offers a wide range of troubling material. Is it effective for parents to simply tell their young children not to explore there? I don't think so. In a recent study, 69 percent of young people admitted to hiding their online activity from their parents.

Having secrets is natural for a child. Snooping and spying on their online lives—pouncing on their unattended cellphone to read their texts or checking their social media activity—is problematic. Studies show that children who have overbearing parents just learn to be more secretive. Worse than that, research shows that when those children run into trouble, the last people they will turn to are their overcontrolling parents. In other words, be vigilant but not a vigilante.

Here's a real-world hypothetical situation: You say to a child, *See that scary house on the hill up there? Don't go there; it's unsafe.* What's the first thing they will want to do?

A decade of online safety messaging may have fundamentally communicated to children that the Internet is unsafe—and therefore super-attractive. The study that looked at the online behavior of children in European countries showed that in just four years between two studies, from 2010 to 2014, as more children were using tablets and computers and became regular users of the Internet, eleven- to sixteen-year-olds were more likely to have received hate messages and been exposed to cyberbullying as well as self-harm and pro-anorexia sites.

This troubling escalation, which could result in more tragedies and harmful experiences online, was mitigated by one very positive change: Children and adolescents in Europe were less likely to have made contact with a person online whom they hadn't met face-to-face. This may show that the various awareness-raising programs and assemblies in schools about meeting strangers in the cyber environment have been effective. A cause for celebration.

The Predator in Your Home

A ten-year-old boy cares what his friends think. Going to sites with violent or sexual content and reporting back to friends what he's seen or heard can increase his reputation among his peers. In a 2012 U.K. report for the National Centre for Prevention of Cruelty to Children (NSPCC), young boys reported that coming to school with "regular" images of hardcore pornography on their phones wasn't cool. But if a boy had an original image—like a picture of his very young girlfriend showing her breasts—that was cool, and so was he.

Young boys generating "original content" . . . This stopped me in my tracks when I read the study. Perhaps the fact that boys are sharing racy photos does not seem like a big deal to you. Boys being boys, right? But, in fact, the behavior reminded me of something far more troubling.

Some years back, when working on the online child-related sex offenders paper, I found that the hierarchical system or social structure of the offenders was very complex. In terms of child-related online sex offenders, the price of entry into higher levels of "closed" child abuse material trading groups was to generate an original indecent image of a child. It proved to other offenders that you weren't a law-enforcement officer. And it proved you had access to a victim—and to original content. Now, very worryingly, it would appear that young boys are beginning to engage in similar behavior.

See how the lines keep moving? See how the cyber norms can shift? If the boys were trading images of Playboy Bunnies, it would seem almost innocent. What's the difference now? They are trading naked and provocative images of real friends, girls from school and the neighborhood. This is children generating and distributing child pornography, albeit of themselves and their peers.

In the real world, children have visible parents, grandparents, caretakers, older siblings, teachers, neighbors, friends, and police to defend and protect their rights. This community of adults will look after a child or children. Very little, if any, of this authority is visible in cyberspace. For a child, it can basically appear that nobody is in charge. And, unfortunately, nobody is.

This becomes problematic when that ten-year-old boy comes home from school and turns on the computer to participate in one of the multiplayer online role-playing games so popular with boys. They are geared for teens and adults, but this makes them seem slightly forbidden, and therefore all the more deliciously exciting, to a younger child. And there are so many games released each year that it can be difficult for parents to stay on top of the content, what's appropriate and what's not.

Multiplayer games have a strong culture—the games can have almost a cult following among fans, to the point of becoming a subculture with its own social rules and slang. A "grind" is a term for any repetitive, time-consuming activity in an online game, and a "griefer" is a player who intentionally provokes anger and irritation, and so on. There are rules of conduct for how to invite new members to join an adventuring party, how to divide the spoils of victory, and how to behave while playing in a new group.

Boys are the predominant participants of multiplayer games due to the focus on shooting and violence, which is particularly attractive to young males. But in the age group of four to twelve years old, a boy can be quite vulnerable, particularly to people who want to take advantage of him. Boys under thirteen routinely play violent shooter games with strangers online—sometimes repeatedly with the same strangers, a percentage of whom are not there just to play.

In my forensic work in the area of cyberbehavioral analysis, one of the things I do is create or build virtual profiles based on available digital evidence. Much like forensic psychologist Georg Sieber and his imagined scenarios of the 1972 Munich Games, I look at what could be possible online. I am a profiler. But the problem is a predator is a profiler too.

And when I use the word *predator,* I am not narrowing this to just pedophiles. It could mean an individual with other motives—profit, extortion, or political radicalization. All of these types of predators have the same modus operandi. They seek out victims, attempt to engage and build trust, and then exploit them.

While a ten-year-old boy is playing his online game, he is hemorrhaging vast amounts of information into cyberspace, even if he's obeying his parents and has not given his name, age, or any other private information to strangers. In exchanges over the Internet, even without use of a webcam, the boy's tone and pitch of voice, as well as use of language, are good indicators of his age. Has his voice deepened yet? Accent and dialect provide more information. The length of time he plays uninterrupted by a parent is another indicator. How late is he allowed to stay up? These hint at how strict or lenient his parents are, at least in terms of the Internet. The pattern of the boy's playing will also tell you the habits or "pattern of life" of his household—when dinner is served, what happens on weekends, what time the parents go to bed. This would help to geolocate the child—that is, find out where he lives.

A predator is hypervigilant, particularly for any information about levels of supervision in that household. The microphone designed in the game to capture the player's verbal input becomes the listening post—in your home—for a predator.

If the predator begins playing regularly with the boy over time, he can figure out where this child lives. Eventually, he can determine or estimate the child's level of social isolation. If a predator is invited into the boy's platoon or group, he can also establish whether the child has friends who consistently join him—another way to determine the boy's level of social isolation. If he's a loner, he could be more vulnerable. The boy's emotional stability can be judged by how he reacts to engineered scenarios in which he is put under pressure during the game—a way to test his resilience. Is the boy upset easily? Is he volatile? Is he reckless? The predator is harvesting data through game play. He will try to figure out if the boy is home alone, or what time of day the parents are likely to be gone. All of this information is available to a potential predator before a one-on-one personal conversation begins with the child.

A young boy can be influenced, brought along slowly, by complimenting him on his exceptional playing style, giving him a support network, and asking him to join a permanent team.

In a multiplayer game online, when more than one predator comes to play a game, sometimes as many as four or five predators, it is called a "wolf pack." While they are circling your child, they will all act like they don't know one another. But they do.

Now imagine if you looked out your window and saw your ten-year-old boy outside playing ball with two or three or four adult men you'd never seen before. Wouldn't you be a little concerned? Wouldn't you go outside to see what was happening—or call your son inside? That's what effectively is happening online. And yet you are less likely to be worried because, after all, your boy is sitting in his bedroom. He's quietly playing his game. He's at home. He's safe. Right?

As a cyberpsychologist, I believe that the lack of supervisory visibility and presence of authority online could be linked to many of the undesirable, inappropriate, and self-destructive behaviors we observe. I would argue that this cyber effect explains a lot of the negative online behavior of children and adolescents.

Law enforcement is aware of the risk in online gaming sites. In the environment of a violent shooter multiplayer game, friendships and loyalties are built over time. One of the powers of the cyber environment is its ability to deceive and delude—and attract vulnerable individuals into strange communities where their desire for acceptance becomes an obsession. Not only do your children encounter strangers online, but they also can find themselves doing things they'd never do in real life, due to the online disinhibition effect.

Children in the eight-to-twelve age group are especially vulnerable to environments where there's a lack of authority. They are just beginning to form their sense of self. They are beginning to bond more closely with peers—while pulling away from their families and their parents. As their friends and schoolmates gain more power and influence over them, a child in this age group feels more pressure to conform. Now take that urge into cyberspace, where behavior of any type can be amplified and escalated. The peer pressure may become greater, even though the other people there aren't physically present. Sometimes they aren't even real.

Murdering to Impress

Morgan Geyser and Anissa Weier of Waukesha, Wisconsin, were ten years old when they became fans of Creepypasta Wiki, a website devoted entirely to horror stories. Kids of almost any age love the paranormal and ghost stories, horror movies, and scaring one another. A *creepy pasta* is a short scary story, like the ones that kids tell one another at a sleepover or around a campfire.

But the immersive nature of the Internet creates a cyber-campfire that is more haunting than a real one. On the Creepypasta site, kids are asked to write their own creepy pasta. The interactivity of posts and comments—the aliveness and immersive nature of the Creepypasta Wiki community—can fire the imagination more powerfully. A warning on the website informs parents that most of the content is fine for kids thirteen and older. But what girl of ten wouldn't want to see something that was deemed fit for one who is thirteen?

Geyser and Weier became enthralled by a menacing phantom called Slender Man and the legend growing up around him. There were scary "photographs" and drawings of Slender Man, and accounts of actual "sightings." Tall and thin, Slender Man wore long black robes similar to an angel of death and was often pictured lurking in forests behind unsuspecting children. Once he'd captured them, he was said to impale them on trees, remove their organs, and drive them to madness. In reality, he was the fabrication of Eric Knudsen, a young husband and father in Florida who had entered a Photoshopping contest on the Something Awful online forum in 2009, hoping to win a prize for designing the best "paranormal image." For the contest, Knudsen created two black-and-white photos of a tall faceless monster in a black suit who stalked children. Eric called him Slender Man. Before long, this character became a focus of fan fiction and a horror hero.

Geyser and Weier believed he was real. They spent afternoons, evenings, and weekends on horror sites, reading posts and comments about Slender Man that described new "sightings." Sometime in early 2014, when they were both twelve, the girls decided to become his "proxies." They believed they would prove their worthiness to Slender Man—and he would appear to them—if they killed someone.

After months of plotting, the girls chose a victim, a schoolmate and twelve-year-old girl who considered herself their friend. Lured into the woods for a "camping trip," she spent a day and a night with Geyser and Weier before the Slender Man proxies attacked her while playing a game of hide-and-seek. She was stabbed in the arms, legs, and torso nineteen times before she was left for dead.

Not quite dead, though, she was able to crawl to a nearby road, where she lay on the sidewalk in her blood-soaked black fleece jacket. A bicyclist found her and called 911.

When arrested, Geyser and Weier confessed in lurid detail, creating a global news story that fueled only more folk-horror interest in Slender Man. As the story in *Newsweek* put it, "Slender Man is now a facsimile of the Puritan devil: He is everywhere, every day, a specter of our anxieties about raising children in a world where technology reigns and the lines between reality and fantasy grow dimmer." This is a real problem for young kids who are still trying to distinguish between reality and fantasy.

How frequently are twelve-year-old girls arrested for attempted murder? In 2012, there were 8,514 people arrested for murder and nonnegligent homicide in the United States. Only one of them was a girl under the age of thirteen.

"This should be a wake-up call for all parents," Waukesha police chief Russell Jack said in a statement given out at a press conference about the stabbing. "The Internet has changed the way we live. It is full of information and wonderful sites that teach and entertain. The Internet can also be full of dark and wicked things."

We read our children fairy tales—sometimes very grotesque ones. One neurobiological explanation for this tradition is to excite pathways of the developing brain through experiences of play and discovery. Fairy tales can also contain important cultural messages about life that are handed down through the generations. Sitting at a campfire, a child can get lost in a scary story and become truly terrified, but soon enough the real world—whether it is the voice of the friend telling the story, or the environment around the child, the smoke from the fire, the darkening sky, the stars above—will remind the child that there is another world and another life beyond the story. But this is much less likely in the cyber environment, where immersion makes fantasies seem real.

Granted, the Slender Man murder attempt is an extreme and thankfully uncommon example, but it illustrates luridly and tragically how individuals can lose sight of the real world in cyberspace. The virtual world can feel as true—or truer. One could potentially lose moral moorings and a sense of right and wrong. By stabbing and killing their friend, the girls apparently hoped to increase their status in the cyber world. But after being tried as adults in a Wisconsin court, they are now prisoners in the real world.

Technically, the girls didn't syndicate online because they already knew each other. But apparently they did self-radicalize, in terms of becoming radical followers of Slender Man together. It is so rare for girls to attempt murder. By talking about their plans for the crime together, and how to do it, the act of murder would have become more normalized between them.

Eventually, Geyser was diagnosed with early-onset schizophrenia during a court-ordered competency evaluation. In March 2016, an HBO documentary, *Beware the Slenderman,* examined both the Wisconsin stabbing and how children can become drawn into beliefs spread via the Internet. To me, this raises a more serious question regarding the role of interactive media and those with a predisposition to psychiatric illnesses such as schizophrenia, which can develop as a result of interplay between biological predisposition and the kind of environment a person is exposed to.

Children are especially susceptible to their environment and can feel pressure to participate in group activities in order to be accepted. Put this together with the known factors of the cyber domain—amplification, escalation, online disinhibition—and the vulnerability of children could become greater, the pressure from peers stronger, and the behavior of the group even more extreme. To not conform may require more confidence, and self-knowledge, and courage, than most eight- to twelve-year-old children have.

The Best Way to Teach Resilience?

The goal of good parenting in Europe, for many years now, has been to teach children resilience. The idea is that by not sheltering and overprotecting them, children are able to find their own footing, strength, and coping skills. This is accomplished by exposing them to difficult realities, hardship, even danger. The approach has gathered quite a lot of support in recent years.

So what is *resilience*?

Resilience is described as "the process of adapting well in the face of adversity, trauma, tragedy, threats, or significant sources of stress—such as family and relationship problems, serious health problems, or workplace and financial stressors." In common parlance, we would call resilience the ability to "bounce back" from difficult experiences, to carry on with a meaningful and productive life in spite of setbacks and even tragedies.

We have all known—or heard inspiring stories about—individuals who recover from awful events and circumstances. And it is always hard to hear about the ones who don't. The stories of individuals in concentration camps in World War II, described so well in neurologist and psychiatrist Viktor Frankl's book *Man's Search for Meaning,* show resilience at its extreme. And the response of many Americans to the September 11, 2001, terrorist attacks—the way the victims' families were able to rebuild their lives and go on, shows a similar, amazing ability to bounce back.

Research demonstrates that resilience is not a trait that people either have or do not have. While there is a biological argument that some individuals are born more resilient than others, the trait is tied to behaviors, thoughts, and actions that can be learned and developed. Studies of children show a gender divide, and an age divide, for resilience.

Boys are especially vulnerable, or less resilient, at nine to ten years. Girls are more vulnerable, or less resilient, when they are a little older—eleven to sixteen years.

How do people learn resilience? There are guides for teachers and parents about how to "teach" resilience—and instill and strengthen the trait. The American Psychological Association has created a list of ten ways to foster more resilience, some of which are just good common sense. The APA advises individuals to make connections and form a community of friends and family who can offer support. It suggests the building of confidence and self-reliance by creating realistic goals and regularly doing things that move you closer to them, taking decisive actions and addressing difficulties that can be improved.

Other suggestions: Avoid seeing setbacks and crises as "insurmountable problems," accept change as a part of life, look for ways to grow and learn from hardships, nurture a more positive view of yourself, keep things in perspective and do not exaggerate problems, remain hopeful and optimistic, take care of yourself by exercising regularly, and do things that are relaxing.

The debate about how best to protect children online has been caught up in, intertwined with, and influenced by the current "teach resilience" philosophy. In 2009, the London School of Economics published a paper by two media and communications professors, Elisabeth Staksrud and Sonia Livingstone, who studied the protection and potential "overprotection" of children online—and how it could result in children who lacked resilience. The authors described "a growing unease that the goal of risk prevention tends to support an overprotective, risk-averse culture that restricts the freedom of online exploration that society encourages for children in other spheres. It is central to adolescence that teenagers learn to anticipate and cope with risk—in short, to become resilient."

The study found that, particularly in Northern European countries with high Internet access, "parental perception of likelihood of online risk to their child is negatively associated with their perceived ability to cope. A comparison of representative surveys conducted among children in three relatively 'high risk' countries (Norway, Ireland and the

United Kingdom) found that although the frequency of exposure to perceived online risks, especially content risks, is fairly high, most children adopt positive (e.g. seek help from friends) or, more commonly, neutral (e.g. ignoring the experience) strategies to cope, although a minority exacerbate the risks (e.g. passing risky content on to friends). Most strategies tend to exclude adult involvement. Significant differences in both risk and coping are found by gender and age across these countries, pointing to different styles of youthful risk management."

Translation: Parents may fret and wring their hands, but the kids are fine—and may even benefit from dealing with content risks online.

I strongly disagree with this approach. Is this how we want to "teach" resilience, by exposing children to extreme content online—to Twitter trolls, hate speech, cyberbullying, and hardcore pornography? This argument suggests that the strides made to protect the innocence of childhood and defend the rights of children in the brutal aftermath of the Industrial Revolution were actually the beginning of a trend toward coddling and hypervigilance. Rather than sending small children into the narrow tunnels of a coal mine or tiny passageways of a tall chimney, we are now sending them into dark, uncharted territory to mine the Web, almost like a character-building exercise.

In the developed world, we historically boycotted goods made by children in Third World sweatshops in the belief that these children were being deprived of the right to childhood. Exposure to extreme disturbing content online also deprives a child of innocence and a childhood. You don't need a rocket scientist or cyberpsychologist to figure that out.

This is the law of the cyber-jungle. This is throw-your-children-into-the-deep-end-of-the-swimming-pool to teach them to swim. If we accept this argument, then it means exposure to pornography is a positive thing—and learning how to fend off adult predators is a skill that should be instilled by experience online. I simply don't agree. Teaching resilience in the cyber context is a white flag of surrender.

The Internet is clearly, unmistakably, and emphatically an adult environment. It simply wasn't designed for children. *So why are they there?* Many experts argue that the positives of the Internet outweigh the damage. If we accept that children are online, will be staying online for greater and greater amounts of their lives, and are by and large having useful and positive experiences there—like learning to read, learning to make friends, and improving fine motor skills and hand-eye coordination—then can we accept responsibility for the damaging things, the disturbing content that could have lasting ill effects on an entire generation?

I would argue that this particular gamble is too great. We cannot gamble with the future development of children who will someday be adults who weren't cared for and raised in the best way. A generation of what I describe as "cyber-feral children." I return to John Suler's analogy: To allow young children online without proper monitoring and supervision is like taking them to New York City to run around, all alone.

And I have to wonder about the children who are falling through the cracks, those who have seen such disturbing things and content that their childhood has been effectively stolen from them. I am thinking of thousands of children in the Philippines who are forced to undress and engage with strange men from all over the world via their webcams. Who is doing more harm to them—the strangers or their own parents, who are selling their innocence online?

Or the children in Africa and other emerging nations who are just now discovering the Internet café. If we think kids in the developing world are vulnerable, what about children elsewhere? Children in Europe and the United States and the rest of the developed world have informed parents, teachers, and other support. What will happen in parts of the world where there is suddenly an increasing availability of devices and services but very little support for these children?

The mean-spiritedness of cyberspace is "fine for an adult," said Andrew Keen, former University of California professor and author of *The Internet Is Not the Answer,* in an RTÉ Radio 1 interview in 2015. "We can deal with it. We have thick skins. But not twelve-year-old girls."

As for childhood? I stand with philosophers Locke and Rousseau in their belief that children have a right to innocence, and a right to a childhood. I think all human beings deserve to have one.

The Rights of a Child in an Age of Technology

Of course, a cellphone in itself is not necessarily dangerous. But now that a greater number of children have access to mobile devices that connect with the Internet, an almost entirely adult environment where we know that one-third of the sites contain pornography, we have to discuss the implications of the impact.

A mobile device offers exposure to inappropriate content—which may lead to the development of addictive or compulsive behaviors. It can expose a child to cyberbullying, or turn a child into a cyberbully. The worst-case scenario involves exposure to forensic risk.

In an age of ubiquitous technology—mobile phones, tablets, and public Wi-Fi—surely monitoring youth behavior online is almost impossible for parents, grandparents, teachers, or caretakers. The point at which technology companies design an app that allows children to take risky images that are transmitted and viewed and then disappear, is the point at which a child's covert behavior is arguably being enabled by technology.

Burdening parents with all the responsibility of cyber-regulations is asking them to raise their families in a lawless environment, a cyber frontier where they must become their own 24/7 sheriff or marshal. Age-inappropriate content is everywhere online, and any tech-savvy child knows how to access it. Have we adequately discussed this as a society? I raised this in an editorial called "Parents Alone Cannot Police Our Youth in Cyberspace" and was drenched afterward with positive responses from thankful parents.

In the real world, we don't expect parents to throw themselves across the doors of all bars and pubs to prevent their children from buying alcohol, or expect them to guard cigarette vending machines to keep them from smoking tobacco. In the real world, kids are kept from buying tickets to movies with sexual and violent content. Printed pornography is kept in special areas of convenience stores.

So why is it so easy to find online?

I argue for more governance, more regulation, and much more protection. For a start, let's look at how the child-protection laws are written now. *Child abuse* is defined by the World Health Organization as "all forms of physical and/or emotional ill-treatment, sexual abuse, neglect or negligent treatment or commercial or other exploitation, resulting in actual or potential harm to the child's health, survival, development or dignity in the context of a relationship of responsibility, trust or power."

When the United Nations set out to describe the rights of children in the late 1980s, prior to the Internet, the result was a document of fifty-four articles and two optional protocols. The basic human rights of children were highlighted: to survival; to develop to the fullest; to protection from harmful influences, abuse, and exploitation; and to participate fully in family, cultural, and social life.

The four main principles of the U.N. Convention on the Rights of the Child are: devotion to the best interests of the child; nondiscrimination; the right to life, survival, and development; and respect for the views of the child. I would argue that many of these articles are being broken by the presence of an unregulated Internet, and that the prevalence of age-inappropriate content online is an international children's-rights issue.

When a child is exposed to extreme content online—be it adult pornography or violence like decapitation, suicide, cutting, or bulimia—who is responsible? Whether it is the device manufacturer, the Internet provider, the host, or the generator of content, I believe all parties involved are collectively participating in the abuse of that child. This should be considered a direct breach of the U.N. Convention, and all those complicit should be held accountable, if not legally, then morally and ethically.

What are the solutions? Age verification and validation online would be a good start. With a little assist from our brilliant tech innovators, age restriction and age verification are possible.

Worldwide, several governments are beginning to try to tackle the current situation—and prepare for a future of more complications. In my opinion one of the world leaders is now Germany, where a long discussion led to the creation of new laws that protect children aged fourteen and under online. The Germans have a word for it, *Selbstgefährdung,* which translates to "self-endangerment" and describes the inevitability of encountering both dangerous content and potential harm in cyber environments. Germany has been working on this since the early 2000s and could be a fine model for the United States and other developed countries that I believe are lagging behind—and have not yet dealt with the matter of the digital age of consent.

In France, the government is looking to tackle extremism on the Internet by legislating that networks that allow hate speech are accomplices in those crimes. In the U.K., there's been a serious debate and now a groundswell of support for more governance. I attended a closed-session hearing at which age-inappropriate content was discussed in the House of Lords, organized by Julia Davidson, a professor of criminology at the University of Middlesex, where I am a research fellow. Now the U.K. government is testing an initiative to require all Internet service providers to offer an effective way to stop adult content from coming into the home. Prime Minister David Cameron stated in 2015 that automatic porn filters, to protect children from the darkness of the Internet, will be made "law of the land." The following year, one of the U.K.'s largest broadband providers, Sky, announced that it will turn on porn filters for all of its new broadband customers—not giving new customers a choice of whether parental controls are on "as standard."

In 2013, I was appointed by the Minister of Communications to the Internet Governance Committee in Ireland and was a relentless royal pain concerning the rights of the child. Shortly after our report was published in 2014, an initiative was launched by Ireland's largest Internet service provider to introduce new opt-in child-safety filters on its home broadband connections as part of an industry initiative to take responsibility for protecting kids from adult and age-inappropriate content. It works on laptops, tablets, and smartphones accessing the Web through the company's service.

Enormously encouraged by this success, I took things one step further in what has become, to be honest, a personal crusade. On the seventieth anniversary of the foundation of the United Nations, I went to The Hague and delivered a keynote address about children's rights in cyberspace. Following that, on Universal Children's Day 2015, I formally proposed that a new amendment be considered to the U.N. Convention on the Rights of the Child that would enshrine children's rights in a cyber context. From my perspective, this is an emergency.

Technology is in a constant state of upgrading, updating, and evolving. Society needs to get much better at keeping up with it. Amending a U.N. Convention is a daunting task and traditionally takes a long time. But in the meanwhile, children are growing up, the content is there—and accessible. A colleague once called me a "disruptive agent of change." But children deserve better. They are our most valuable resource and our hope for the future, as U.S. president John F. Kennedy said. We all need to get a little less complacent and a lot more disruptive.

CHAPTER 5

Teenagers, Monkeys, and Mirrors

Inside her classroom at Coral Springs Charter High School, Susana Halleck was in distress. The Florida teacher, seven months pregnant, was suddenly experiencing labor contractions. She sat down in a desk chair and struggled to endure the pain—her mouth open, her eyes wide, one hand on her brow. That's when one of her students, junior Malik Whiter, class of 2015, pulled out his mobile phone.

It was time for a selfie.

In dreads, cap, and big sunglasses, Whiter flashed a big, happy-go-lucky grin for his camera while angling the lens to show his grimacing, pain-stricken teacher in the background. "Selfie with my teacher while she having contractions," he tweeted.

Selfies bring new meaning to the word *self-conscious*. These quick, seemingly innocent self-portraits—typically taken with a smartphone or webcam and shared via social media—serve many functions. They can be a preened vision of a public self, a bragging moment of accomplishment, a display of humor, or a declaration of irony to the world, almost a performance. The ubiquitous mobile phone with its mirror-image camera technology makes self-portraits easy to take, delete, filter, or fix, and even easier to share.

Some kids would call what Malik Whiter did, taking his own picture with a featured but unsuspecting person in the background, a kind of photobomb selfie. It's a prank or joke. Photobomb moments are something like a tourist's snapshot souvenirs. *I was there.* But this time, the background wasn't Mount Rushmore or Niagara Falls. It was Halleck's suffering.

Whatever you call it, in the time it took for Halleck to reach the hospital to be examined by doctors, Whiter's pic was making the rounds on social media, first to other high school students at Coral Springs, and then quickly beyond. By evening, it was viral—and had been retweeted by thousands. When asked later by local TV news reporters what possessed him, Whiter said he was just hoping to record the unexpected event for himself and "for her." He had asked Halleck to smile for the camera, he said, but when she refused, he had no other choice but to catch her "off guard."

It went viral mainly because people found it funny. BuzzFeed raved: "Behold! The greatest selfie of all time." Was it funny? Sure, if you don't take a moment to consider this act in a deeper way—and what it means to use a human being in distress as a visual joke in the background of a curated self-portrait shared on a public social network.

There are more troubling trends to notice here—invasion of privacy, breach of good manners, absence of empathy, not to mention a demonstrated lack of respect for pregnancy, motherhood, the classroom setting, and a teacher's authority. I could continue this list for another page. But let's be honest with ourselves: Nobody looks to teenagers as role models of civility and decorum. They can be jokesters, disrupters, provocateurs. Pushing the limits is what they do best. Why? In psychology we explain that they are forming *self-concept,* or identity, and enjoy experimenting with boundaries and taking risks.

They also crave feedback, which helps them figure out, eventually, who they are—and what the world expects of them. So when teenagers take selfies and share them, what are they hoping to discover? Probably themselves.

Prior to the Internet, this crucial time of identity formation was spent in the real world—a more intimate greenhouse where feedback, both positive and negative, was received from a real-world audience of friends, family, and figures of authority. The social norms and what was expected of these developing human beings was fairly consistent. Twenty or thirty years ago, would a teenager have been allowed to take a photograph of a distressed teacher in a classroom and, without permission, been allowed to publish it in a magazine?

As for Whiter, his teacher's early labor was declared "a false alarm," and she returned to the classroom two days later. By then, her image had become an online sensation, or "meme." (It has since been retweeted 60,424 times and favorited 64,808 times.) Had Whiter crossed a line? Did he get in trouble? Halleck didn't mind, says Whiter (who got a B in her class). "I have honestly laughed until I cried looking at that pic on two different occasions tonight," he tweeted. "Think I'm making it my wallpaper." By the month's end—a long time in cyberspace—the picture had been appropriated by other pranksters online who cropped Whiter's smiling face from the classroom setting and superimposed him in the foreground of historical events like the crucifixion of Christ and the *Hindenburg* blimp explosion.

The Internet is now a primary adventure zone where teenagers interact, play, socialize, learn, experiment, take risks—and eventually figure out who they are. This chapter will try to grapple with this shift, and look at the impact of this new environment on youthful identity formation.

Could growing up in cyberspace change a teenager's sense of self?

Why So Heartless, Selfie?

The same year as Malik Whiter's cyber-celebrity moment, another controversial selfie was seen by millions. A lovely young woman with long blond hair, aviator sunglasses, white knit scarf, and matching hat was caught in the act of posing for her own selfie while, behind her, a suicidal man was hanging on the rails of the Brooklyn Bridge.

What, aside from basic psychopathic tendencies, would cause a person to be so cold and unfeeling about another human being's emotional crisis? Let's stop and contemplate this. Just as Whiter made a joke of his teacher's moment of physical crisis, the young blond (she would remain anonymous), whether she planned to share her selfie with a wide audience or not, was apparently making fun of a stranger who was so emotionally troubled and confused that he wanted to end his life. Yes, her selfie seems more heartless than Whiter's selfie, but aren't the sense of disengagement and lack of empathy eerily similar? I am not the only person to notice this. The day after the Brooklyn Bridge moment, an observer's photograph of the anonymous young woman took over the entire front page of the *New York Post* with the apt headline "SELFIE-ISH."

This slap of disapproval only encouraged a new trend. (You know how *that* goes.) In 2014, when traffic was stopped on a Los Angeles freeway due to a man threatening to jump from an overpass, a group of drivers left their cars to pose—big smiles—for group shots and selfies with the suicidal man in the distance behind them. The same year, a policeman in Istanbul was called to the scene of the Bosporus Bridge, where a desperate individual was clinging to the rails. The suicidal man jumped three hours later, but before he went, the officer took a selfie. The bridge and the jumper were in the background. More recently, in March 2016—in perhaps the ultimate example of this trend—a hostage on an EgyptAir flight posed for a bizarre smiling selfie next to a hijacker in his suicide vest. (If you google "suicidal" and "selfie," you can find more of these.)

Let's try to consider the mind-set of these people—not the distressed suicidal individuals, but the selfie-takers.

Were they conscious of what they were doing? Or were they so lost, so separated from ethics and empathy, that they weren't able to clearly consider their actions? Are they emotionally impaired, or has cyberspace impacted their judgment?

A condition that results in lack of empathy toward another person's distress is *narcissism*. This is a personality trait that exists to varying degrees in almost all human beings and can be facilitated by cyberspace. I'll have more to say about this later in the chapter, but for now, I want to acknowledge that a little narcissism can be a good thing, psychologically speaking, and even necessary. It can be one of the drivers behind achievement—leading individuals to seek attention, acclaim, fame, prizes, or special treatment. Actors are famously perceived as the ultimate narcissists, and the psychologically healthier ones even crack jokes about it. They aren't necessarily heartless people. But a narcissist's desire to be noticed and become a focus of attention can override a concern for other people—and result in callousness about their suffering.

As with so many personality traits, psychologists have defined a spectrum of narcissism—generally assessed by the Narcissistic Personality Inventory (NPI). Individuals with high scores demonstrate an inflated sense of their own importance, grandiosity, extreme selfishness, enormous self-regard, and a deep need for admiration. Behind the mask of ultraconfidence, their self-esteem is very fragile and vulnerable to criticism.

Why get into all this?

Because teenagers (as well as children) can display narcissistic-type traits due to the simple fact that their sense of self, or "self-concept," is still being formed. They can seem to be uncaring about others because they are distracted by the work of creating an identity. Teenagers will try on new selves and new clothes and new hairstyles to the point of total disengagement with anything else going on in their family life or home. For a teenager, this sort of experimentation, along with risk-taking, is one way that identity is formed. Going too far is part of the process—almost a requirement.

Who am I today? Who do I want to be tomorrow morning? They look for answers in the feedback they receive from their peers. And today, to a greater and greater degree, this feedback happens online, not just from their friends but in free online astrological profiles, personality questionnaires, and a plethora of phone apps that will analyze their handwriting, music tastes, food preferences, and even bathing styles. (My favorite: an app that analyzes your personality based on how you take a shower.)

This incessant need for personality feedback is what fuels the popularity of annoyingly ubiquitous online tests, like BuzzFeed's "Which Disney character are you?" and the online permutations of the Myers-Briggs personality inventory, now loosely applied to an array of scientifically dubious outcomes, like "Which Personality Type Makes the Most Money?"

Teenagers are consumed by their own reflections, in other words, hoping to figure out who they are. What happens when the bathroom mirror, where teens used to stare at themselves, is replaced with a virtual mirror—a selfie that they just took with their phone?

Monkeys and Mirrors

In a famous study done forty years ago, great apes—chimpanzees, orangutans, and gorillas—born in the wild were placed before a full-length mirror on a wall. At first, the wild chimps reacted as if another chimp had appeared in the room; they vocalized and made other threatening gestures at the mirror. After two or three days, they began to understand the image in the mirror as a reflection of themselves in some way. Interestingly, they began exploring their own bodies before the mirror—studying parts of themselves that they hadn't seen before, or couldn't see without the use of a mirror.

In psychology, one way to describe what happens in front of a mirror is called *mirror-image stimulation,* referring specifically to "a situation in which an organism is confronted with its own reflection in a mirror." An animal that shows signs of recognizing the image in the mirror as its own is said to have "passed the mirror test," which is strong evidence of having developed *self-concept*. This is not innate, but learned.

Self-concept is used in human social psychology to describe how people think about, evaluate, or perceive themselves. The actual definition is "the individual's belief about himself or herself, including the person's attributes and who and what the self is." A monkey that has self-concept demonstrates an awareness of a self that is separate and distinct from others, as well as constant.

Early socialization has been shown to impact a monkey's behavior in front of a mirror—and presumably its development of self-concept. In a comparative study of rhesus monkeys born in the wild and rhesus monkeys raised in captivity or isolation ("isolates"), there were distinct differences in mirror studies. Wild monkeys raised around other wild monkeys, and therefore socialized, grew uninterested in the mirror eventually—and returned to a live group of fellow monkeys for physical interaction. But monkeys raised in isolation did not. The isolates remained fascinated with their own reflections to the point of disinterest in others.

This study may tell us something important about children and teenagers, since looking at your own image on the screen of a mobile phone is a form of mirror-image stimulation. I know I'm comparing teenagers to monkeys, but bear with me.

What are teenagers learning about themselves by looking at their own selfies? Could this impact the development of self-concept? The study also raises this question: Could young people who have grown up with too much technology and not enough face-to-face interaction with peers remain more isolated, like the monkey isolates, retreating to the comfort of their own digital reflection rather than turning to their friends or family for comfort and physical interaction?

Could this cyber effect encourage children or young teenagers to lose interest in others—or never develop it in the first place?

Since there hasn't been time for proper developmental studies in this area, we just don't know. But in the case of many human and primate studies, the similarities are worth noting—and it would be a mistake to dismiss them. When humans are born with visual defects and then undergo operations that give them sight for the first time, their initial response to a mirror is the same as a monkey's initial response: They believe they are looking at another person, not themselves. Human babies respond similarly, first seeing their reflection in a mirror as a playmate. Most children don't show signs of self-recognition until they are approaching two years of age.

Humanistic psychologist Carl Rogers's work is valuable in terms of illustrating how a young person develops identity. He described self-concept as having three components:

1. The view you have of yourself—or "self-image."
2. How much value you place on your worth—or "self-esteem."
3. What you wish you were like—or "the ideal self."

I think we should consider adding a fourth aspect of "self" to Rogers's list. In the age of technology, identity appears to be increasingly developed through the gateway of a different self, a less tangible one, a digital creation.

Let's call this "the cyber self"—or who you are in a digital context. This is the idealized self, the person you wish to be, and therefore an important aspect of self-concept. It is a potential new you that now manifests in a new environment, cyberspace. To an increasing extent, it is the virtual self that today's teenager is busy assembling, creating, and experimenting with. Each year, as technology becomes a more dominant factor in the lives of teens, the cyber self is what interacts with others, needs a bigger time investment, and has the promise of becoming an overnight viral celebrity. The selfie is the frontline cyber self, a highly manipulated artifact that has been created and curated for public consumption.

But how do we explain that weird, vacant, unmistakable expression on the faces of many selfie subjects? The eyes look out but the mind is elsewhere.

The virtual mirror could be socially isolating, except for one thing. The selfie can't exist in a vacuum. The selfie needs feedback. A cyberpsychologist might say that's the whole point of a selfie.

Selfies ask a question of their audience: *Like me like this?*

The Psychology of Feedback

To understand feedback more deeply, we need to go way back to the work of sociologist Charles Horton Cooley in 1900, decades before the advent of the Internet or when monkeys were stuck in front of mirrors. Cooley came up with what he called *the looking-glass theory.* (*Looking glass* is an archaic English term for mirror.) Cooley used the concept of a person studying his or her own reflection as a way to describe how individuals come to see or know themselves.

In the case of Cooley's looking glass, the information that we use to learn about ourselves isn't provided in a mirror's reflection. It is provided by others—their comments about us, the way they treat us, and things they say. In the looking-glass self, a person views himself or herself through others' eyes and in turn gains identity. In other words, he argued, the human self-concept was dependent upon social feedback.

Philosopher William James, the so-called father of psychology, expanded this idea by pointing out that individuals become different people, and express their identity in different ways, depending on whom they are with. He argued that we could potentially have as many "selves" as we have friends, family members, and colleagues who know us.

Now let's fast-forward to the next century and do the math—and consider the psychology of this effect in cyberspace. If you have a repertoire of many selves—potentially as many as people who know you—social media could exponentially expand the number of selves you create. Is your "self" environment-specific? Are you the same person on Twitter, Facebook, Instagram, WhatsApp, Snapchat, and LinkedIn? Does this new explosion of selves cause a splintering of identity or, particularly for young teenagers who are going through critical stages of identity formation, cause developmental problems? And what about critical feedback?

Presenting yourself to the world is a risky business. It's hard to imagine an individual alive who hasn't experienced some form of rejection, subtle or strong, embarrassing or humiliating. But you can also be accepted for the self you present—and feel rewarded by pleasant feelings of pride and affection.

Let's imagine you have just turned thirteen. The five years ahead of you are a natural time for questioning and seeking. You'll be trying new clothes, mannerisms, friends, interests, and pastimes. You'll probably begin experimenting with what you think of as adult behavior. This helps you make sense of the self within, as you unconsciously piece together an identity, like a collage. You are working to create a constant, steady, reliable, knowable, and familiar self.

What kind of information—or feedback—is the virtual mirror going to give you? In this regard, the cyber environment may be much more overwhelming than the real world. To begin with, the sheer number of "friends" has grown, and therefore the volume of feedback will be far greater. Prior to the Internet, a teenager would have a limited number of social groups to juggle—family and extended family, schoolmates, maybe neighbors. Now the number of social groups is potentially limitless. How would you begin to interpret, filter, and process that vast amount of information coming from so many different quadrants?

Imagine how confusing and potentially unsettling it could be. A virtual tsunami of feedback to process.

Now factor this in: The cyber self is always under construction, psychologically and digitally. Even while the real you is sleeping, the cyber you continues to exist. It is "always on"—evolving, updating, making friends, making connections, gaining followers, getting "likes," and being tagged. This can create a feeling of urgency, a continuous feedback loop, a sense of needing to invest more and more time in order to keep the virtual self current, relevant, and popular.

This may explain the obsessive interest among teens in curating their selfies. When the process of identity formation in real life becomes confusing and difficult to control, as it is for most teenagers at some point, what could be more satisfying than being able to perfectly calibrate and manage the portrait that the online world sees? To some extent, we all engage in image management, but it now begins at an earlier age, and in some cases before identity has been properly formed. This may lead to identity confusion.

After all, which matters the most: your real-world self or the one you've created online?

Probably the one with greater visibility.

Narcissism

Narcissism, a tendency found in teenagers during their identity-formation years, is a term taken from Greek mythology, after Narcissus, a very handsome young man with a fantastic physique. One day he wanders into the woods and discovers a pond. When he leans down to drink water, he sees his own reflection and falls in love with himself. Realizing that the love he seeks is impossible, he despairs. Unable to leave the pond and his own reflection, he dies.

An individual with extreme narcissism demonstrates preoccupation with self and how he or she is perceived by others. Lack of empathy and selfishness are both traits. A normal developing child displays what looks like narcissistic and self-centered behavior. A toddler who will not share toys is self-centered. A child who thinks the world revolves around them is also rarely interested in the adults around, or doesn't show curiosity or empathy for others. For them, others don't really exist, except to serve the needs of the child.

There's even a study by Piaget to prove this, the famous *three mountains task*. In the study, a child is taken to stand on one side of a table where a model of a mountain range has been placed. On the opposite side of the table, a doll sits in a chair. The child is shown sets of drawings and asked to pick out the scene that shows the mountain scene the way the child sees it. Nearly all kids can do this without difficulty.

The child is then asked to pick out the drawing that shows how the doll sees the mountain scene. Most children can't do this. They will pick the drawing that shows their own view. This experiment is used to demonstrate how young children have difficulty imagining the perspective of others.

Over time, self-centeredness lessens. By preadolescence, children have learned to be more other-directed, courteous, kind, and empathetic. They have been reminded—by example, by parents, and by other figures of authority—that others exist in the world and others matter.

Once they are teenagers, though, they can become more self-centered again. This comes with their developmental preoccupation with creating their identity. Erik Erikson described this period of development between ages twelve and eighteen as a stage of *identity versus role confusion,* when individuals become fascinated with their appearance because their bodies and faces are changing so dramatically. Once this stage of development is finished, the self becomes reintegrated—and wants to act in the world, and get out of "self." The discomfort about appearance fades away. Usually by the late teens, individuals become more self-accepting.

Narcissistic behavior, in other words, is considered a natural part of development and is usually outgrown. But according to recent studies of young adults, it would seem that fewer of them are moving on beyond their narcissistic behavior. A study of U.S. college students found a significant increase in scores on the Narcissistic Personality Inventory between 1982 and 2006. It's pretty interesting that this change coincides with the rise of the Internet.

Of course, there could be lots of contributing factors for the rise in narcissism. It could be caused by a generation of indulgent parents who have overpraised their children. (A study done by Ohio State University suggests that constantly praising kids for their tiniest of accomplishments may have the unintended side effect of developing an overinflated ego.) The predominance of narcissism could have occurred alongside the rise of a more individualistic society. The "dark side" of self-reliance and individualism is an increase in self-regard.

Social-networking sites encourage the sharing of personal information, which can result in a preoccupation with self. Certainly, there has never been a time in history when people were expected to show off their accomplishments, their travel photos, their new clothes and hairstyles, and every restaurant meal, to such an extent. The social pressure for young people is tremendous to create and share photos—and participate in this hyper-celebration of self. It's interesting to think about how humans have colonized the Internet and turned it from a platform of sharing into a platform for self-promotion.

Some amount of narcissism is considered healthy, as I've said. But I wouldn't want to overstate how healthy. There is a line between feeling good about yourself and grandiosity, which can interfere with relationships and become dysfunctional. Being high on the narcissism spectrum is a tough, usually problematic way to go through life. Like the eponymous Narcissus, securing a lasting and rewarding relationship with another human being often eludes these individuals. Sometimes they resort to marrying another narcissist and forming a "mutual admiration society" (the marriage in the Netflix show *House of Cards* is a perfect example of this), although these relationships are often volatile and rarely last.

I will explore this a bit more in the next chapter, about cyber-relationships, but the term "narcissistic relationship" may qualify as an oxymoron. Let's hope today's teenagers aren't on this path. But since they've been hanging out in cyberspace during their developmental years, it is too early to tell.

Filtering Your Ideal Self

In Oscar Wilde's philosophical novel *The Picture of Dorian Gray*, an impossibly beautiful young man, Dorian Gray, has his full-length portrait painted in oil by a famous artist. The portrait is so lovely, it saddens Dorian to think that in his real life, day by day, he will age and become an old man while the portrait remains spellbinding. It will stay perfect and always remind him of what he's lost. Devastated by this thought, he makes a wish—something like a deal with the devil. He wishes that the painting would grow old instead of him.

As the years pass, Dorian descends into hedonism and immoral behavior to the point of depravity (the kind of stuff I covered in chapter 1, "The Normalization of a Fetish," which you may still be trying to get out of your head). But his wish comes true. He never appears to age. Only his portrait shows the ravages of time and the effects of his corruption. You would think this might be an ideal situation, but his eternally young and beautiful appearance leaves him isolated from real life—from love, the effects of time, and eventually morality. His portrait grows uglier and more sinister every day, reflecting Dorian's inner reality. The painting becomes so horrific that he decides to put it in a locked room where nobody can see it.

Thanks to technology, Dorian's story can now happen in reverse: The cyber self gets better with each year—filtered, Photoshopped, and, most important, "liked"—while the real self can be parked in a locked room. The beauty industry has always relied on the insecurities and perfectionism of young women for a sizable percentage of its annual profits. Now there's a plethora of apps designed to meet the same need, altering not the physical body but the virtual representation of it online. An app called PhotoWonder (with 100 million users in 218 countries) can make you appear slimmer and also give you a much-needed tan. With a little help from Facetune, you can appear taller and with blemish-free skin. The CreamCam app improves your skin tone, removes blemishes and oil "shine," and can erase any flyaway hair around your face. And let's not forget Skinny Camera, which supplies an instant virtual weight loss of ten to twenty pounds. Anybody can be a supermodel online.

A subtler but no-less-telling aspect of curating a selfie and conveying status online is clothing. What a teenager wears, or doesn't wear, is full of information. There's an old saying that the human being is composed of three parts: soul, body, and clothes. More than one hundred years ago, William James observed that this was more than just a joke—that the relationship between clothing and the self is intimate and deep. And let's not forget Oscar Wilde, who said, "It's superficial not to judge by appearances." In forensic psychology, clothing is especially important, as it, or lack of it, is considered "behavioral evidence of intent." In other words, clothing is an indication of what the wearer is about to do. When a woman is found dead in the mountains and she's wearing high heels, you know she didn't go there to hike.

Human beings dress with an outcome in mind, even if subconsciously. For a teenager, clothing is an intense and complicated affair. Teenagers are keenly sensitive about what they are wearing and how they present themselves, and use clothing to try on roles and identity, still feeling uncertain about where they may land. They are also enormously influenced by friends and popular culture. Teens want to look cool and knowing, perhaps often in direct proportion to how uncool and unknowing they feel inside.

It's a cliché that some adolescent and teenage girls like to wear provocative clothing in order to look more grown-up. But in recent years, the trend toward sexually provocative clothing in this age group has become more extreme. I can't help but think the Internet, and its effect of quickly normalizing extremes, has something to do with this. As fashion online becomes more revealing and more provocative, and as sexts become more prevalent (half the teenagers in the United States report having sent or received one), I believe that the clothing we see on teenagers in the real world has been impacted. Of course, what is "provocative" is subjective.

Trying to talk about this subject always reminds me of discussions of global warming. The average temperature across the planet may be increasing, but since it is happening incrementally—and we may be getting used to these changes, to the point of its seeming normal—it is hard to notice. But when you hear the words "last year was the hottest on record," it is meaningful. And I think it's fair to say, without question, that the clothing of teenagers today is more sexually provocative than ever before. How do we judge what is appropriately "sexy" and what is inappropriately "too sexy for a young teenager"?

Reality check: Should we even be using the words *sexy* and *provocative* in the same sentence as "young teenager"?

I was recently doing a forensic cyberpsychology study and needed to assess a sample of digital images to analyze a number of behavioral factors; one of these was whether the clothing worn by young subjects in certain photographs was "sexually provocative" or not. After looking at one hundred images, I had to give up. It became clear to me, as I tried to scale the provocativeness levels of the supershort shorts, miniskirts, high heels, bare tummies, Playboy Bunny logos, and thongs on eleven-year-old girls, that I had researcher bias. What does that mean? It means that you are not able to study something in an objective way. Simply put, I find the trend of online posing by and provocative clothing on adolescent and young teenagers a worrying development. I am puzzled by the social pressure that girls are under to skip adolescence and go from ten to eighteen almost overnight. They have become increasingly consumed with making themselves sexually attractive to boys, just as young boys are consumed with attempts to present themselves as hypermasculine. This premature sexualization is happening before children and young people are capable of dealing with it emotionally, mentally, or physically.

Teenage sexuality has always been a challenging construct to society, and every generation seems to complain that its teenagers are growing up too quickly. "Youth is subjected by our civilization to aggressive sex stimuli and suggestiveness oozing from every pore," declared the education professor Clark Hetherington in 1914, who condemned the proliferation of racy movies and tell-all magazines.

While I am not advocating a moral panic regarding the phenomenon of sexualized teens, there is a whole new layer to it when we consider cyber effects. Setting aside developmental risks, in terms of identity formation and self-concept, there is also a problem with the images left behind. While provocative selfies are intended for peers, the reality is far more troubling. They now feed the deviant interests of adults and offenders who are sexually attracted to young teens. These images are hacked and stolen from social media and have ended up on pornography sites. In some cases, young teens have located their own images on these sites, read the lewd and disturbing comments, and been traumatized. Worrying about contemporary sexualization of teenagers is one thing. But worrying about actual forensic risk is another.

Cyber-Migration

Plastic surgery among teenagers is another area that has been impacted by the norms online. The easy curating of selfies may be linked to a rise in plastic surgery. According to a 2014 study by the American Academy of Facial Plastic and Reconstructive Surgery (AAFPRS), more than half of the facial surgeons polled reported an increase in cosmetic surgery for people under thirty. There is also a rise in children and teenagers requesting teeth whitening and veneers reported at dental clinics. "Social platforms like Instagram, Snapchat, and the iPhone app Selfie .im . . . force patients to hold a digital microscope up to their own image and often look at it with a more self-critical eye than ever before," explains Dr. Edward Farrior, president of the AAFPRS. "These images are often the first impressions young people put out there to prospective friends, romantic interests, and employers, and our patients want to put their best face forward."

Sadly, surgeons have reported that bullying is also a cause of children and teens asking for plastic surgery, usually as a result of being bullied rather than a way to prevent it.

Okay, so let's put all these trends and technological developments together—from teenagers using apps to filter and "improve" the appearance of their selfies to the rise of plastic surgery among young people, the escalation in provocative self-presentation, and the quest for the perfect body. What do they tell us, given that we know that human beings look to feedback in order to develop identity?

Imagine for a moment the shy thirteen-year-old who feels uncomfortable speaking to others. For this child, posting a selfie will be easier and more rewarding—no actual contact! Now imagine that child progressing through the stages of identity formation and never having to practice being a human being on the stage of real life. This is what causes isolation in adulthood.

Now consider this phenomenon as it may impact slightly less shy children—ones who, like many adolescents, find it stressful or difficult to express themselves in a real-world setting and prefer to retreat to the computer screen for social contact. There, it is easier to be the exciting, attractive person they wish to be. Why would you want your real self visible in the real world if your cyber self is so much better? But this creates a problem down the road.

Which is the ideal self?

And which self is in the locked room?

This matter is complicated by the legitimacy of the virtual world. Much of our lives—for adults as well as teens—now occurs online. How we present ourselves socially matters, which is the central idea behind the socioeconomic term *social capital*—meaning that our social networks have actual economic value. In other words, who we know and how we look could have an impact on our earning potential. Online, social capital can be important, even for teenagers, as they hear about schools and find employment and internships, and can earn real money on social-network sites with product endorsement and ad monetization. This simply makes their online presence more credible and important, and adds to the time and energy that goes into the commodification of self. (Sure, build a social network, but be careful, you may not get that internship if your Facebook wall is smothered in your own selfies.)

Things don't end well for Dorian Gray. In every way, his existence is corrupt. Decades of pleasure-seeking and debauchery bring him, finally, to a moment of truth and self-loathing. In a fit of anger, he kills the artist who painted his portrait; later he slashes the painting, trying to destroy the evidence of his guilty conscience. Dorian is found dead on the ground below the portrait, which has been magically restored to its youthful, beautiful version, while his own body—deformed, ugly—now reflects the true state of his soul.

In our contemporary metaphor, it is the cyber self that does not age—but can potentially live an idealized life forever online. But there is another self that is aging and human. This is the self that needs to be cared for, respected, nurtured, and nourished.

This is the self that needs the most love.

But if it's kept in the locked room, how will love find it? And can you love *that self* as much as you love your selfie?

Body Dysmorphic Disorder

The story of Tallulah Willis illustrates how a teenager's life can spiral out of control when the cyber self doesn't seem to measure up and isn't accepted. The daughter of actors Bruce Willis and Demi Moore spent her early years idyllically on a ranch in Idaho, offline and away from the spotlight. It wasn't until she was in third grade, when her family moved to Los Angeles, that she discovered how truly famous her parents were and what that meant for her—in terms of reflected glare and heightened expectations. She began to feel unworthy and, by comparison, unspecial.

In a candid personal essay in *Teen Vogue,* she traces the moment her life began a sad descent into self-hate and drugs. She was thirteen, sitting in a New York hotel room, looking at a photograph of herself online:

> I broke down in tears as I started to read the comments. I thought, I am a hideous, disgusting-looking person. I might be nice and I might be kind, but I'm a really unattractive human being. In that moment, a switch flipped. It wasn't about the anonymous cyberbullies—I became my own worst critic.
>
> In high school I continually searched for something that would make me feel OK. I felt like I had to dress really skanky and be the loud, stupid, drunk girl at the party. I figured people would like that, and I was beating them to the punch.
>
> Then I started to get boobs, and that's when my eating disorder really sparked. I began starving myself. I was trying to become the superquiet girl who smoked a lot of weed and was really skinny and serious. In the same way I used substances, I used shopping and even social media for vapid validation. I would dye my hair and get a tattoo or piercing. There was always something that I believed would fix things, some way I could avoid being myself.

By the time Willis arrived at college, her depression was overwhelming. She couldn't sleep, wouldn't talk to anyone, hated eating, and felt "so removed from my body and from my mind that it was like I was living in a cardboard replica of what life should be." Her sister Scout came to her rescue, and "forced me to see what I was doing. There wasn't a huge, horrible moment, but I knew I needed to go take care of myself." She went to inpatient treatment for forty-five days and faced her substance abuse and self-hatred. What she realized seems simple, but it's a lesson that bears repeating.

> While in treatment, I thought a lot about my 5-year-old self. I was a towhead, and my hair stuck out in all directions, like I just got electrically shocked. I had tiny Chiclet gremlin teeth, and I was always running around like a ragamuffin. I realized that I love little me; I adore her and I want to protect her. Then I realized that I don't love big me and I haven't ever loved big me. When I look at old pictures, I remember that that little person is inside me all the time—and I need to be the person to take care of her now.

For Tallulah Willis to reconcile the cyber self who was attacked, she dug deep and went back to her authentic, younger, real-world self and tried to reconnect. It took courage to tell her story so openly, which is undoubtedly inspirational for many.

The struggle with self-image, self-esteem, and self-love doesn't just affect teenage girls. Boys can suffer just as much from confusion during the years of identity formation, and flounder—turning to alcohol, drugs, and other intoxicants to numb the sense of being out of control. And they can become just as fixated on appearance.

In 2015, a significant rise in the number of boys and young men suffering from eating disorders—which includes anorexia, bulimia, and binge eating—was reported in Ireland and the U.K., a jump of as much as 30 percent since 2000, a problem that is linked to concern for having a "gym body" and looking buff. It is not a coincidence that this echoes the rise of social media, or the proliferation of those full-length mirror shots posed as selfies. "In the gym, emphasis is not on exercising for fun or sport, it is actually on exercising in front of a mirror to become bigger," said Dr. Terence Larkin, consulting psychiatrist at Saint John of God Hospital in South Dublin. "Men are beginning to go down that road of overvaluing physical appearance, and self-esteem is becoming more and more dependent on that."

Take Danny Bowman, a teen in Britain who aspired to a life of Internet celebrity. He was fifteen when he began posting selfies on Facebook, not unlike millions of teens across the world. Seems innocent enough, right? But he became increasingly focused on his appearance, how good he looked in his selfies, and he began dreaming about becoming a model. Judging from his selfies, he felt confident that he had a shot. When a modeling agency turned him down, he was devastated. By the age of seventeen, he was a self-proclaimed "selfie addict" who had lost interest in school, friends, and sports. Perfecting his selfies was his passion, his cyber self a work of art.

The problem was his real self. By the age of nineteen, Bowman was unable to leave the house and spent up to ten hours a day preoccupied with taking his own photograph with his phone, trying to improve on his self-portrait. "I was constantly in search of taking the perfect selfie," Bowman told the *Mirror*, a British newspaper, "and when I realized I couldn't, I wanted to die." In frustration and despair, he took an overdose—but fortunately, he was found in time by his mother.

Doctors prescribed intensive therapy for Bowman, who was treated for technology addiction, OCD, and body dysmorphic disorder (BDD), a form of anxiety that causes sufferers to worry excessively about their appearance—a condition that apparently has become more common in the era of the selfie. I believe this condition could be aggravated or facilitated by the virtual mirror, so it is worth discussing in a bit more detail.

At some point in their development, many teenagers come to suspect that there is something wrong with their physical appearance. They want to look like their best friend, or they want to look like an actor they admire. They have idealized role models and feel their own self is falling short. They may make declarations like "I hate my nose" or "My legs are too skinny." Adults should be reassuring, supportive, and kind. And if siblings are incapable of being supportive and complimentary, they should remain quiet.

Everyone has a self-perceived flaw, or two, or three. None of us is born perfect—or stays perfect over time. Most people come to accept their appearance the way they come to accept other things in life that are beyond their control. What's the point in obsessing over a mole or wrinkle, or the way your ears stick out? You might make a solid effort to disguise the flaw or wear things that distract from it. Then you move forward with more important matters.

But individuals with body dysmorphic disorder are obsessed with imagined or minor defects, and this belief can severely impair their lives and cause distress. Here is a list of some common signs and symptoms, as offered by the Mayo Clinic:

- Preoccupation with your physical appearance with extreme self-consciousness
- Frequent examination of yourself in the mirror, or the opposite, avoidance of mirrors altogether
- Strong belief that you have an abnormality or defect in your appearance that makes you ugly
- Belief that others take special notice of your appearance in a negative way
- Avoidance of social situations
- Feeling the need to stay housebound
- The need to seek reassurance about your appearance from others
- Frequent cosmetic procedures with little satisfaction
- Comparison of your appearance with that of others

You may obsess over any part of your body, and which physical feature you focus on may change over time. The features most commonly obsessed over include:

- Face, such as nose, complexion, wrinkles, acne, and other blemishes
- Hair, such as appearance, thinning, and baldness
- Skin and vein appearance
- Breast size
- Muscle size and tone

- Genitalia
- Teeth

As with other true psychological conditions and disorders, the amount of distress is a defining factor. Individuals with BDD are completely convinced that there's something wrong with their appearance—and no matter how reassuring friends and family, or even plastic surgeons, can be, they cannot be dissuaded. In some cases, they can be reluctant to seek help, due to extreme and painful self-consciousness. But if left untreated, BDD does not often improve or resolve itself, but may become worse over time—and, as in the case of Danny Bowman, can lead to suicidal thoughts or suicidal behavior.

Self-Actualization

In her compelling story, Tallulah Willis described feeling detached from her true self: *so removed from my body and from my mind that it was like I was living in a cardboard replica of what life should be.*

I would argue that the cyber self, while it offers glimpses into who you are, is a literally detached self. This cyber self is like a hand puppet that is speaking for you but isn't really you—and can actually be quite different from the authentic real-world you. In other words, the real you has turned the cyber you into an object: The selfie is proof of this objectification. By posting a selfie, you are required to experience yourself as an object that is presentable or not. You judge your selfie from a detached distance, even if it is posted impulsively.

I would argue that this self-objectification, and the sense of detachment from true self, could explain many of the negative behaviors seen online and discussed in earlier chapters. It feeds disassociation. Detached from your cyber self, you can feel detached from your actions—and come to believe you aren't truly accountable. Now let's think about a teenager in the process of identity formation from the age of ten, eleven, twelve to late teens, a crucial window of time to create a strong foundation and sense of self. This process is critical to development, and can have an enormous impact on the rest of an individual's life and sense of self-esteem.

But wait. There's another new complexity. Instead of one solid identity to create and accept, there are now two—the real self and the cyber one.

Carl Rogers described "self-actualization" as an ongoing process of always striving to be one's ideal self. A "self-actualized" person is one whose "ideal self" is *congruent,* or the same as his or her perceived actual self or self-image. Rogers believed that this sense of being, or having become, the person you want to be is a good marker for happiness, and a sign of a fully functioning individual. If you accept his description of happiness, then it's troubling to see the results of a survey of children and teens, ages eleven to sixteen, in which half agreed with this statement: "I find it easier to be myself on the Internet than when I am with people face-to-face."

The transition from childhood to adulthood is a critical developmental phase, what psychologist Erik Erikson described as a "psychosocial stage." For an awkward adolescent or teen, it may be a lot easier to avoid painful experiences performed on the stage of real life, but these are often important developmental milestones and come with consequences if missed. Identity may not be fully developed—and what one wants to do or "be," in terms of a future adult role, may not be fully explored. Social coping skills may not be acquired. Learning to navigate the tension or lack of comfort that the real world sometimes brings is necessary for the developmental process, as youth explore possibilities and begin to form their own identity based on their explorations.

Failure to successfully complete a psychosocial stage can also result in a reduced ability to complete further stages. For Erikson, the next stage is *intimacy versus isolation,* occurring between ages eighteen and forty, when individuals learn how to share more intimately with others and explore relationships that lead toward long-term commitments with someone other than a family member. Avoiding intimacy or fearing relationships or commitment can lead to isolation, loneliness, and often depression.

This is why we need to talk more about the repercussions of teenagers failing to establish a sense of identity in real life as well as cyber life. The result of such a failure can be what Erikson calls "role confusion," when young people become unsure about themselves or their place in society. Stanford University psychologist Philip Zimbardo believes that contemporary boys are in crisis due to excessive use of technology. As he writes: "The digital self becomes less and less like the real-life operator."

The cyber self is a masterful creation—funnier, wittier, better looking than the real self. But the problem lies with the vulnerability of this split-self existence. And it's a serious problem. Here's what I mean: If you look at all the studies done over the past ten years on cyberbullying, you'll see that few of the solutions and awareness campaigns have worked effectively. This is why I am working hard to create and employ an algorithm to target it. Each year, more and more teenagers are devastated, even destroyed, by experiencing bullying online. Why?

Think of the time and energy that teenagers put into their cyber selves—the self-portraits they've painted. When the cyber self is attacked—called "stupid," "ugly," "a loner," "a loser"—then, I believe, this could cause a catastrophic interpsychic conflict, an emotional clash of opposing impulses within oneself.

If the best version of you that technology can produce is rejected, how does that make you feel about the only self that's left, your real one? There are no studies in this area, but I believe it is critical to explore if technology-facilitated interpsychic conflict could lead to self-harm.

Sext + Sensibility

There are loads of scary stories about sexting, but I'll start with a relatively unscary one. The story dates way, way back to the summer of 2007, a million years ago in the sexting debate time line. But it marks a turning point.

Two thirteen-year-old girls from northeastern Pennsylvania, Marissa Miller and Grace Kelly, were hanging out together and goofing around on a hot muggy night when they decided to strip down to their white sports bras and underwear.

They hammed it up—Kelly flashing a peace sign—while a third friend took a photograph of the two girls.

This photo was shared and shared again. Eventually it traveled all the way to the administrator's office of Tunkhannock Area High School, where the girls were both students. It was found when school officials confiscated five cellphones of other students. Boys at the school had been trading photos of semi-dressed, semi-nude, or totally nude teenage girls. When a local prosecutor threatened to charge Miller and Kelly effectively with the generation and distribution of child pornography unless they agreed to attend a five-week after-school program, the girls and their families protested—saying the picture was taken in innocent fun, and was never meant to be distributed. The intention behind the act of taking the photograph was hardly the same as the intention behind the generation of child abuse material, but the law does not yet make this distinction.

"There was absolutely nothing wrong with that photograph," said Marissa's mother, MaryJo. "I certainly don't want pedophiles looking at my daughter in her bra, but I don't think that was the intention to begin with. This is absolutely wrong. . . . It's abuse of his authority."

The American Civil Liberties Union (ACLU) asked a federal judge to block the D.A. from filing charges—stating that the teens didn't consent to having the picture distributed and that the image was not pornography, in any event.

While some may consider sexting to be only a social concern, the legal implications are clear. According to U.S. legal scholar Mary Graw Leary, "the production and dissemination of self-produced child pornography is, under the law, the production and dissemination of illegal child pornography." Attempts at prosecution for sexting haven't been confined to the United States. In Australia, child pornography laws have been enforced in connection with sexting, and between 2009 and 2011 more than 450 teenagers were arrested and charged with circulation of indecent images of minors.

Prosecution is not a deterrent. The incidence of sexting has only risen. In 2008, the same year that Miller and Kelly were threatened with charges, a U.S. study reported that 39 percent of teens had taken a sexually charged or explicit image of themselves with their mobile phones and sent it via text message. Six years later, the numbers had grown to include half of all teenagers having sent a sext, according to a survey of undergraduates at Drexel University. Can we really try to outlaw—and describe as immoral and indecent—a practice that now involves half the population of teenagers in the United States?

Sexual curiosity is a natural part of being a teen. From strip poker to spin the bottle, teens have been doing this for generations. Back in the day, they went behind the cowshed and flashed each other. The difference was, nobody captured digital evidence and widely distributed it.

"I'm not saying that it's healthy or that it's harmless, but it's not a situation where kids who are depressed are doing this or kids who have very bad self-esteem are doing it," said Elizabeth Englander, a professor of psychology at Bridgewater State University in Massachusetts, in response to the Drexel study. "It's engaged in by many kids who are functioning well and not having problems."

Translation: Sexting is a new cyber norm.

What do I think about this? As with any new and still-evolving behavior, I think a balanced approach is the route to the best outcome. But the debate itself—and the gray moral middle ground of sexting—just highlights, once again, the vulnerability of teenagers and younger adolescents in the cyber environment. And once again, courts, school administrators, parents, and child psychologists are desperately playing catch-up. But to stay ahead of the new problems emerging in cyberspace would require adapting our laws and installing new education programs at the same warp speed that new technology is developed and marketed, which is a practical impossibility.

The competing views on the issue are fascinating. In their seminal article "Sext and Sensibility," David Finkelhor and Janis Wolak of the Crimes Against Children Research Center at the University of New Hampshire argue that school officials, law enforcement, parents, and legislators have overreacted to the phenomenon of teenage sexting—and could do more harm by exaggerating the actual problems and then responding inappropriately. Finkelhor and Wolak caution against the "big stampede" to get educational programs into schools to warn kids to stop before they destroy their lives or end up in jail—or on a sex offenders registry. Scare tactics, they believe, are often the worst way to communicate a message to teenagers and have been shown to fail miserably "whether we're trying to warn kids about the dire consequences of drug use, premarital sex or delinquent behavior." Worst of all, it could have "boomerang effects" and result in reluctance to come forward out of fear of prosecution.

Since the numbers of sexted images that actually show naked genitalia or breasts are minimal, Finkelhor and Wolak argue that defining these images as pornography is alarmist. "The child porn issue," they say, "is really clouding our thinking about this problem."

The authors call for a moratorium on any sexting-education programs until the right strategy can be conceptualized. In my opinion, a good approach would be incorporating information about sexual images and sexting into the curricula of general sex education—which deals with bullying, dating, and interpersonal relationships. (And I hear this is now being done in some school systems.) The content needs to be developed in close cooperation with teenagers, and the message needs to be evaluated for effectiveness. Just telling teens not to do something isn't enough. We need to understand the motive and psychology of the behavior as well as teach kids about legal consequences.

The current laws do not deal adequately with the problem of sexting. At Middlesex University I have conducted extensive research in this area—involving the London Metropolitan Police, the Los Angeles Police Department, and the Australian Federal Police under the umbrella of the INTERPOL Specialists Group on Crime Against Children. Following four years of research on three continents, I came to the conclusion that one of the problems is that sexting is viewed almost exclusively through the legal lens of child pornography, and while it is true that the images can look very similar, they are very different in terms of intent. In one instance, the explicit image taken by a teenage girlfriend and boyfriend is done voluntarily, and at the other end of the spectrum, the image is coerced from a child victim by a criminal sex offender. There needs to be an active review of the law in this area and the creation of a legal classification framework that differentiates between teenage voluntary sexual exploration and criminal generation of child abuse material, the defining criteria being *mens rea,* or intent. I am actively working on proposed amendments to legislation in this area through my involvement with The Hague Justice Portal in Europe.

Do I agree with Finkelhor and Wolak that there's been an overreaction to the rise of sexting? No. As a society, we need to pay attention to cyber effects, the ongoing evolution of behavior, and, importantly, how we react, adapt, and respond. It is impossible for me to ignore the stories of truly vulnerable individuals who have been tragically impacted by the behavior.

The darkest place where the practice of sexting can take a teenager is actually part of my work. The risks of any online behavior need to be discussed carefully and thoroughly rather than swept under the rug. We have to make sure we've educated kids properly and created adequate measures to protect them.

Revenge + Sextortion

Jesse Logan was a vivacious, artistic girl, a junior in high school in Ohio, whose boyfriend asked her to send him nude pictures of herself. After Logan and her boyfriend broke up, he shared the nude images with younger girls at the high school, who started calling Logan a "slut" and a "whore."

Miserable and ashamed, Logan began missing class—a cry for help, and a sign of a teenager in crisis.

Her mom, Cynthia Logan, didn't discover that her daughter was cutting classes until she was notified by the school. When Jesse told her about the photographs—and the bullies—Cynthia Logan pressed the school to take strong action. But when administrators didn't do enough to satisfy her, Cynthia convinced her daughter to appear on a local Cincinnati TV station and tell her story. That was May 2008.

"I just want to make sure," Jesse told the interviewer, "that no one else will have to go through this again." Two months later, she hanged herself in her bedroom. She was eighteen.

The point of this story isn't just to shock and scare you. What it tells me is that how sexting is handled—by parents, by the media, by schools and courts—may be more potentially risky than the behavior itself. Jesse's tragic suicide brings me to another important point. The number of teens—and even younger children—who are engaging in self-harm is now more evident than ever. This urgently needs to be researched and investigated. I believe self-destructive behavior is amplified and even encouraged by information readily available on sites and forums that target vulnerable youth—the worst cyber effect of all.

There's a name for the sharing of indecent images by an embittered ex-boyfriend or ex-girlfriend as a way to "get back": *revenge porn.* It's a burgeoning trend and discussed in the media. But is it appropriate to call it "porn"? As I've mentioned, the intention behind the taking of these images is very different from marketed pornography—and calling it "revenge porn" just seems to further victimize a victim. But whatever you call it, there are many efforts now under way to try to combat the escalation of this behavior. But until recently, before it was given a name by the media, authorities had a hard time figuring out how to talk about it, or act on it.

Jesse Logan's story resonates with another well-known case, that of Amanda Todd, a fifteen-year-old girl in Canada whose circulated sexual image led to her becoming a target of online bullying in 2012. Teenagers usually don't have the resilience and strong sense of self in place to handle the fallout from being stuck in the role of a whistle-blower or an anti-sexting missionary. Why put this responsibility on a vulnerable child who has already been a victim? An investigation of Todd's hard drive provided more details. It wasn't a shared sext that got her into trouble, but an indecent image. Todd had been in a webcam chat room with more than 150 people when she decided to lift her shirt and flash her breasts to the camera.

It was an impulsive act, but one that she was unable to leave behind. Someone took a screen capture of the image and sent the link to all of Todd's Facebook friends. That's how the image of Todd's flashing moment fell into the hands of online predators, individuals who troll the Web looking for embarrassing pictures, like hers, and then contact the subjects for *sextortion* purposes, or blackmailing, typically by threatening to post the explicit images more widely on the Internet.

Victims of sextortion are sometimes asked for money, but more often they are asked for more photographs—or to perform sex acts for the camera. Todd had been approached by sinister blackmailers, as the disturbing records of her chat log revealed:

> [I am] the guy who last year made you change school. Got your door kicked in by the cops—give me 3 shows and I will disappear forever . . . if you go to a new school, new bf, new friends, new whatever, I will be there again, I am crazy yes.

Hindsight is 20/20, and maybe vulnerable Amanda Todd should have stopped having an online presence at this point. Instead, she created a haunting video and posted it on YouTube. Without showing her face, she told her moving story of depression, anxiety, panic disorder, and self-harm in cue cards. One month later, she killed herself.

Two years later, a thirty-five-year-old man living alone in an isolated resort town in the Netherlands was charged with extortion, Internet luring, criminal harassment, and child pornography in the Amanda Todd case. He was suspected of numerous other instances of online abuse in the Netherlands, the U.K., and the U.S.

Let's backtrack a bit. What was behind Todd's original act—which led to her tragic ending? Impulsivity. It is certainly a psychological factor that contributes to the production of a sext, but it is also linked to immaturity. Young people have less self-control and have a much harder time resisting an impulsive urge. Their desire to explore overrides the risks of impulsivity.

How many teenagers have mooned a passing car on the highway or flipped the bird at a police car or other figure of authority—and paid a price for it? Many. And they went on to live perfectly decent, law-abiding lives. I am not saying these dumb mistakes of adolescence should be encouraged, but I believe they should be accepted as part of a long process called *growing up*. Mistakes are supposed to be made. And teenagers will make them. In the case of Amanda Todd, she lifted her shirt and flashed her breasts. Should it have been the end of the world for her?

There's one more important point that needs to be made as I finish retelling the sad story of Amanda Todd. Her video has 19 million views now and is still available online. I have to weigh in—and share my opinion of "memorial sites" on Facebook that honor teenagers who have died or killed themselves.

I realize that this is where family members visit and leave updates and posts, and express their sorrow and remember their departed loved ones. And yes, it's true that families and friends may find great comfort in these sites. But I have a greater concern. This type of memorial afterlife is very powerful—and can appear to leave a legacy, and bestow fame, on the departed teen. I am concerned that this could lead to more self-harm and more suicide attempts. Some teens fantasize about dying as a way to get attention or even revenge. The Internet has the power to turn that need into a spectacle. I believe we need to reevaluate the constructiveness and impact of the memorial sites, and consider taking them down.

There should be no upside to suicide, in the real world or in the cyber one.

The Privacy Paradox

In real life, would a teenage girl walk around with a photograph of herself naked—and show everybody at school? Would she undress in class and pose suggestively? That's what happens, potentially, every time a sext is sent. Besides impulsivity and narcissism, what are the other possible explanations for this disinhibited behavioral shift online?

It could be this simple: It's fun.

You probably don't need to be told that kids like to have fun. But for scientific purposes, it has been demonstrated in studies that having fun is another important part of development. In psychology, it's generally referred to as *play*. Could sexting constitute an *adaptive form of play*? Psychologists haul out the word *adaptive* whenever they want to describe behavior that is changing in order to keep up with either the environment or evolving social mores. Considering that as many as half of all U.S. teens report sexting, it could simply be another way of "keeping up."

Another way of saying this: *Mom, all the other kids are sexting*. But what about privacy? Don't teenagers worry about their photos careening around cyberspace and being seen by strangers? This brings up another fascinating and much explored area of research. In cyberpsychology it is called the *privacy paradox,* a theory that was developed by

Susan B. Barnes, a professor at the Rochester Institute of Technology, to demonstrate how teenagers exhibit a lack of concern about their privacy online. It's an interesting shift because so often in the real world, many teens are self-conscious and tend to seek privacy. But online, something else happens. Even teenagers who are well versed in the dangers and have read stories of identity theft, sextortion, cyberbullying, cybercrimes, and worse continue to share as though there is no risk.

In the early days of the Internet, this paradox was somewhat understandable. Very few people were able to imagine how behavior would shift and escalate online. In 2005, when the Facebook accounts of four thousand students were studied, it was discovered that only a small percentage had changed the default privacy settings. The most recent study, done in 2015, shows that now 55 percent of teenagers in the United States have adjusted their Facebook settings to restrict total strangers from viewing their content. While that shows a change to greater concern about privacy, it still is too low a number.

The explanation is lack of interest.

Teenagers simply don't care.

Why?

Because privacy is a generational construct. It means one thing to baby boomers, something else to millennials, and a completely different thing to today's teenagers. (In ancient Greece, it meant nothing at all; there wasn't even a word for "privacy.") So when we talk about "privacy" concerns on the Internet, it would be helpful if we were talking about the same thing—but we aren't.

But just because teenagers don't have the same concerns about privacy as their parents do—and don't care who knows their age, religion, location, or shopping habits—it doesn't mean they don't pay attention to who is seeing their posts and pictures.

According to danah boyd, the TED Talk celebrity and visiting professor at New York University, most teenagers scrutinize what they post online very carefully. In her book *It's Complicated: The Social Lives of Networked Teens,* she argues that teens adjust what they present online depending on the audience they want to impress. Everything is calibrated for a specific purpose—to look cool, or tough, or hot.

In other words, when it suits them, teenagers can be enormously savvy about how to protect the things they want kept private, mostly from their parents. "The kind of privacy adolescents want is the same kind of privacy that they have always wanted," says Ian Miller, who studies the psychology of online sharing at the University of Toronto. "They don't care if Facebook knows their religion, but they do care if their parents find out about their sex life."

For my generation, a "secret" is something you tell only your closest friend—and swear them to an oath for life. For teenagers, a "secret" is shared online with four hundred of their closest friends, some of whom they've never met face-to-face.

This explains why the "privacy" debate has been muddled. There's a communication breakdown. Although we share the same language, we are sharing a label that refers to two very different things. How about a new label—*open privacy*—a different concept from "privacy," since teens have a different understanding of what they feel is appropriate to share with friends, friends of friends, and basically strangers.

The Risky Shift

We've explored explanations for why an individual teenager might post a heartless selfie and why teens don't consider their privacy in the same ways that adults do. But why does sexting continue to be popular with teenagers when it has been shown repeatedly to be a bad idea? And why has it become more prevalent, in spite of all the school lectures, social awareness campaigns, and so forth?

Group dynamics are a powerful force. Individuals behave differently when part of a group than when they are alone. For example, it has been proven that teenagers in particular can be judgment-impaired when in a group of peers, due to what's called the *risky-shift phenomenon.* (And this is why in many jurisdictions in the United States, teenage drivers are restricted from riding in cars with more than one other teenager.) It was James A. F. Stoner who first discovered the tendency in groups toward riskier behavior. Stoner found that each participant's opinion became more extreme as a result of the group conversation.

Groupthink is another way to look at how an individual's behavior changes in a group. The larger the group, the more people tend to conform. Logically, the same thing is true online.

So it goes without saying that large groups of teens online, connected by social networks, are likely to behave in riskier ways, and they will also feel more peer pressure the larger their online social group is.

Consider the story of the "sexting ring" that was uncovered at Cañon City High School in Colorado. In November 2015, students in a small town of sixteen thousand were found to be circulating between three hundred and four hundred nude images of classmates as well as some eighth graders in the local middle school. Some of the photos were even taken on school grounds, and then uploaded onto a shared site.

The students used a "ghost app" that they'd downloaded to their mobile phones as a vault to store these images and then trade them like baseball or Pokémon cards. A ghost app hides itself on the screen of a device by appearing to be a calculator or something equally innocuous, which keeps parents or other authorities from noticing. And as with Snapchat, images can be permanent or created to disappear.

An undisclosed number of students were suspended for their participation in the sexting ring, including members of the football team—which led to the cancellation of a game. The alleged crimes, possession and distribution of child pornography, are both felonies. But due to the age of the students, who are all minors, local law enforcement wasn't sure how to respond. Said George Welsh, the school district's superintendent, "There isn't a school in the United States probably at this point that hasn't at some point dealt with the issue of sexting."

What would make hundreds of small-town kids behave in such a way? If you combine the effects of increased sexualization and attention-seeking, obsession with cyber self, new definitions of "privacy," conformity and groupthink enhanced by the strong ties of social-networking connections, and a *cyber-risky-shift phenomenon,* the previously unthinkable becomes possible.

When Pimps Go Online

Cutting-edge journalism is more necessary than ever, as the speed of change in the cyber world outpaces academic studies. We have to thank

Wired magazine for breaking the story of the indictment of Marvin Chavelle Epps, one of a growing number of sexual traffickers who use the Internet to find new young recruits and rent out their services in adult ads. To meet the supply and demand of the sex trade, there's nothing as fast and easy as cyberspace.

The beginning of *Wired*'s 2009 story "Pimps Go Online" should send a flash of fear into the heart of any parent of a teenager:

> She was a 16-year-old California girl looking for trouble on MySpace; he was a 22-year-old self-described pimp who liked the revealing photos she posted to her profile. Three weeks after they met on the social networking site, they were arrested together in real life outside a cheap motel in Sacramento, 50 miles from her home. She was turning tricks. On her arm, a fresh tattoo showed bundles of cash and her new acquaintance's street moniker in 72-point cursive.

Epps was a "new kind of pimp," a tech- and Web-savvy guy who used sites like Craigslist to prostitute girls. On his chat log online, he even gave out free business advice to other potential pimps: "Get some professional, beautiful, elegant, glamor shots, put 'em on these escort websites, and her phone gone slap."

Here's the forensics piece: You may say to yourself that the risk of your own daughter meeting a pimp online is small—and that could be true. But the demographic is shifting. "We're seeing kids who are getting into this stuff that do not match society's stereotype," Ernie Allen, president and CEO of the National Center for Missing and Exploited Children, told *Wired*. "These are not just kids in poor families who have no other options. We're seeing kids from the full spectrum of society, and a lot of that is due to recruitment over the Internet."

Girls who were already at risk of recruitment into prostitution now face escalated circumstances and danger. They may be acting out. They may be looking for money. Already at a vulnerable age, they can be easy to find—and preyed upon. On social-networking sites, these girls give cues that they are vulnerable. And pimps have ways of knowing how to exploit them.

Asia Graves is a case in point. The daughter of a drug addict, Graves began working as a prostitute on the streets of inner-city Boston at the age of sixteen to keep her family afloat. She was picked up by her first trafficker in the middle of a snowstorm. He told Graves that he'd help her, take care of her. "He showed me the ropes," she told *USA Today* in an interview in 2012. "If we didn't call him Daddy, he would slap us, beat us, choke us."

Sold from one trafficker to another, Graves was forced to sleep with men in New York, Atlanta, Philadelphia, Atlantic City, and Miami. She posed for Craigslist and Backpage.com ads and set up "dates." She worked six days a week for up to $2,500 a night.

"You think what you're doing is right when you're in that lifestyle," Graves said. "You drink alcohol to ease the stress. Red Bulls kept you awake, and cigarettes kept you from being hungry."

I had a chance to meet Asia Graves in 2012, when I was recruited to work on a White House research initiative to find technological solutions to technology-facilitated human trafficking. Asia was a participant: A beautiful, intelligent, poised young woman, she silenced an entire room of power-suited White House working-group members as she told her moving story.

The phrase *human trafficking* is a broad umbrella. It is sometimes just called "trafficking in persons." Either way, it is a term that describes any instance when one person obtains or holds another person in compelled service. The U.S. Department of State Report in 2011 described major forms of trafficking as forced labor, sex trafficking, bonded labor, debt bondage among migrant laborers, involuntary domestic servitude, forced child labor, child soldiers, and child sex trafficking. It has been proven that technology has created a new dimension for these cybercriminals in which to operate, and has changed the way they do business.

On the White House project, I was introduced to and worked closely with Dr. Steve Chan, a big data scientist, in a research group that used machine intelligence to analyze huge volumes of online images publicly posted on escort sites. Our research was designed to help law enforcement identify young trafficking victims online.

In one month alone—April 2013—the leading U.S. publisher of online prostitution published 67,800 listings for "escorts" and "body rubs," both considered euphemisms for "prostitution"—in twenty-three U.S. cities. But prostitution can also be a proxy for human trafficking, and, in yet another cyber evolution, what was once visible on street corners and in hotel bars is now an increasingly invisible online activity.

According to a 2015 police report, it has been established that online classified adult advertising facilitates trafficking. The sheer volume of online classified adult advertising and potential human trafficking activity, along with the increasingly tech-savvy cybercriminal, is resulting in a problem of epic proportions. Human trafficking is not confined to a specific geographic locale. With more than 20 million victims of human trafficking around the world, it has become an issue of global importance.

We published our paper for this work in 2015. Using a corpus of publicly available big data, we were able to move toward big insights and contribute to scientific knowledge in this area. Indeed, the implications of our research may extend far beyond the problems posed by human trafficking and into the realm of tackling other technology-facilitated exploitative crimes, such as the generation and distribution of child abuse material.

Courage and resolve propelled Asia Graves to finally escape from the control of traffickers. She joined a group of former prostitutes who had founded FAIR Girls (which stands for Free, Aware, Inspired, and Restored), a girl-empowerment and anti–human trafficking organization based in Washington, D.C. They were behind the development of an app called the Charm Alarm, which girls on the street can use to call for help. FAIR Girls has also been instrumental in getting trafficking sites shut down.

Craigslist was dubbed "the Walmart of child sex trafficking" and vilified for years for allegedly fostering sexual abuse. But it finally shut down its "adult services" section in September 2010. Backpage picked up where Craigslist left off, making an estimated $22.7 million annually from thousands of ads for young women—many of them teenagers—claiming to be escorts, strippers, or massage therapists. They are photographed wearing very little clothing and in suggestive poses. The good news is that advocates like FAIR Girls study the advertisements and reach out directly to offer victims jobs, housing, legal support, medical treatment, and a new life. In late 2015, the U.S. Senate voted unanimously to file civil contempt proceedings against Backpage. The action followed the site's refusal to attend a Senate hearing to provide information regarding online child sex trafficking. This is the first time in more than twenty years that the Senate has taken such extraordinary action. I'm delighted to see a proactive stance on a contemporary cyber problem.

I + Me

We owe teenagers an apology. That's what I think. We are failing to protect and defend them in cyberspace. We are failing to understand and therefore protect their developing selves. Tech companies have made billions of dollars while looking the other way. Opportunistically, they have jumped in to offer solutions to emerging obstacles, creating social platforms such as Snapchat, Wickr, Confide, and the German-based Sicher, where risqué images can be sent and viewed. While they supposedly can disappear almost as soon as they are posted, in fact there are many ways they can be saved. I have serious issues with the facilitation of covert and potentially illegal behavior of minors. I find it inherently unethical.

Do teenagers need to explore and have adventures? Yes. And we should let them. But the risks in the cyber environment are real: sextortion, predation, cyberbullying, and harassment.

And what about the more nuanced and much-harder-to-study risk of harm to a developing identity? Juggling two selves, the real one and the cyber one, is a lot to expect of young individuals who are still figuring things out, about themselves and the world. We are likely a decade away from seeing the cyber effect on psychological and emotional well-being and the formation of a sturdy and sustainable self. We can see signs and clues coming already, I believe, in the new norms of sexting, the obsession with the cyber self, premature sexualization, the plastic surgery among younger people, the escalation of body and eating disorders, and the rise of narcissistic behavior (if not true narcissistic personality disorder). These trends should be cause for alarm. Narcissism and excessive self-involvement are both known attributes of those who suffer chronic unhappiness.

Eighty years ago, the American philosopher and social psychologist George Herbert Mead had something very relevant to say about how we think about ourselves—and express who we are. Like William James before him, Mead studied the use of first-person pronouns as a basis for describing the process of self-reflection. How we use *I* and *me* demonstrates how we think of self and identity.

A child learning English has to master the distinction between *I* and *me* and know when to use them.

The self-referential pronoun *I* is appropriate only when speaking for yourself, from yourself, delivering authentic expressions subjectively. The objective pronoun *me* is appropriate only when the child is talking about himself or herself as the object. Mead showed that these two pronouns point to two complementary views that each person has of the self.

There is *I*.

And there is *me*.

By using *I*, a child shows that he or she has a conscious understanding of the self on some level. When using *I*, the child speaks directly from that self. The use of *me* requires the child to have an understanding of himself or herself as a social object. To use *me* means figuratively leaving one's body and being a separate object.

Where is the *I* in contemporary times? It is the authentic subjective and conscious real-world self. The selfie, the frontline cyberspace expression of the cyber self, is all *me*. It is an object—a social artifact that has no deep layer. This may explain why the expressions on the faces of selfie subjects seem so empty. There is no consciousness. The digital photo is a superficial cyber self.

It's also interesting to note how *me* as a self-referential pronoun has expanded its hold on language, despite grammatical rules otherwise. The overuse of *me* has spread like wildfire. Teenagers, teachers, even broadcasters on TV, can be heard saying, "Me and my BFF" or "Me and my mom . . ."

A teenager may think he or she is creating a better "self," a better object, with each selfie. But I would argue that every selfie taken, and improved upon, causes an erosion or dismissal of the true self. With each selfie taken, and invested in, the true self is diminished. In a way, it reminds me of the aboriginal cultures that believe each portrait photograph robs the soul. And when I think of cyberbullying and read accounts of the self-harm associated with it, I can't help but wonder if the cyber self isn't punishing the real one.

In the late nineteenth century, the pioneering American psychologist G. Stanley Hall, sometimes called "the father of adolescence," argued that the teenage years are a natural period of *Sturm und Drang,* or "storm and stress," during which kids will often experience mood swings, fight with parents, and engage in risky or dangerous behavior. We can't blame the Internet for that. But we can wake up and see that it's even more important to protect them there.

I think it's finally time for what I call *pro-techno-social initiatives*—that is, technology developments for the greater good. It is time for the tech industry to step up and pay attention to social problems associated with use of its products. Mark Zuckerberg and his wife, Priscilla Chan, have pledged to donate 99 percent of their Facebook shares to the cause of human advancement. That represents about $45 billion at Facebook's current valuation. I would respectfully suggest that all of this money be directed toward human problems associated with social media.

There are also things that parents can do—starting today. To begin with, remember that it is your child's real self that wants and needs to be loved, accepted, and nurtured. Do what you can to pull kids back to *I* and not let them drift to *me.* This is strengthened by conversation.

- Ask them about their real-world day, and don't forget to ask them about what's happening in their cyber life.
- Tell them about risks in the real world, accompanied by real stories—then tell them about evolving risks online, and how to not show vulnerability.
- Talk about identity formation and what it means—distinguishing between the real-world self and the cyber self.

- Talk about body dysmorphia, eating disorders, body image, and self-esteem—and the ways their technology use may not be constructive.
- Tell your girls not to allow themselves to become a sex object—and tell your boys not to treat girls as objects online—or anywhere else.

And, parents of teenagers, if you find a sext, sit down and talk about it. Resist the urge to shut down or confiscate all your son's or daughter's devices. The point at which you banish your teenager to his or her bedroom—hating themselves, hating you, and hating their lives—and take away their phone and computer, you are depriving them of their entire support system. That can be too hard. They need to vent. They need to reach out to friends. Let them. And finally, if anything goes wrong in their cyber life, tell them not to try to handle it on their own. That's what parents of teenagers are for.

CHAPTER 6

Cyber Romance

One of the timeless principles of all great love stories is that intimacy provides the spark of love, but separation is what causes it to ignite and become a blaze. Consider how the classic romances play out, whether it is Tristan and Isolde, Romeo and Juliet, Scarlett and Rhett, or Rick and Ilsa. Ties between two individuals are created in small and subtle moments as they come to know and trust each other. Romantic love and passion are fanned and fueled when obstacles are encountered.

Obstacles are romantic, in other words. They can be situational—misunderstandings, mistaken identities, bad timing, physical upheaval, prior commitments, lack of reciprocity, geographical separation, or unavailability due to marital status. The importance of obstacles explains why "playing hard to get" can escalate the emotion of the other partner in an affair and why, for some, a romantic prospect who is unavailable, physically or emotionally, can provoke the most urgent feelings.

Obstacles and intimacy work in tandem, a push and pull that preoccupy two people as they head toward an intimate union of some kind, whether it is a missed opportunity, a sexual encounter, a walk down the aisle, or something in between. A succinct definition of *intimacy* is "close familiarity or friendship," but it is often a euphemism for "sexual intercourse." For psychologists, it is a complex construct particularly at a point in time when human-to-human contact and the courtship process are increasingly mediated by technology.

How has intimacy—or sex—been changed? Couples find and sustain relationships in different ways—often facilitated by webcams, proximity apps, texting, sexting, and social media. But has this made things better, more satisfying, more rewarding? This chapter looks at the ways cyber effects are shifting mating rituals and romance. And the story has a twist: Even though love turns out to be easier to find online—just a swipe right away—is it really love?

Stranger on the Train Syndrome

Way back in 1996, at the dawn of the Internet age, communications expert Joseph Walther coined the term *hyperpersonal interaction* to describe the way individuals tend to communicate online. According to Walther, the reason that things turn intimate so quickly online is due to the lack of visual cues, which is another way of talking about the effects of *invisibility*. Since two people who meet online are initially unable to see each other in most cases, their relationship begins differently. Invisibility has a substantial impact. While online, you are hidden—either partially or completely—particularly if you are communicating on a visually lean platform without Skype or live streaming—and therefore trust is created almost solely by self-disclosure.

Trust is essential to intimacy. In the real world, we know how to assess the trustworthiness of other people. Everything from a handshake to facial expression, eye contact, tone of voice, posture, and language come into play and inform our overall opinion of a person. We might approach a person we find attractive and test the water with jokes or flattery. We learn quickly, by his or her reaction, if they are interested in more.

On traditional dating sites like Match.com and eHarmony, this meet-and-greet process is initially more laborious. You are required to fill out a form and supply personal information, details such as job, age, birthplace, education level, income range, and even your favorite books and movies. After creating a personal profile on a dating site, there is a quiet period of mutual examination as individuals who have been introduced to each other by the algorithm decide whether they find each other's profile photographs attractive (a very important part of mating). If there are "sparks," things begin to take off.

And "sparks" is a good way to describe it. Many significant regions of the human brain literally light up on MRI scans when an individual is experiencing feelings of romantic love. Chinese researchers studying the "science of love" have found that the left dorsal anterior cingulate cortex (dACC) becomes active, as well as the insula, caudate, amygdala, nucleus accumbens, temporo-parietal junction, posterior cingulate cortex, medial prefrontal cortex, inferior parietal lobule, precuneus, and temporal lobe.

Pretty impressive! Although I have to acknowledge this is probably more so to a scientist than a die-hard romantic.

Online, the courtship ritual continues in an exchange of texts, email, or chat-room posts, each more intimate than the last—which can make cyberspace dating feel like a journey to the confessional box. This is because in order to establish trust online—and form a bond—you need to say more, and reveal more, and describe more. In many ways this is standard in traditional dating; once it is decided that two people might be interested in each other, they try to establish ties and trust, and (like all love affairs) there is the added value of getting a chance to self-reflect and see yourself through another person's eyes. People who fall in love don't just learn a lot about another person; they learn a lot about themselves, in the form of feedback. We look out into the eyes of another and see our own images reflected back. In psychology, this is known as Cooley's classic looking-glass self, as discussed in chapter 5.

But there's one huge difference: Traditional dating relationships can take months to develop to the same level of intimacy that is reached almost instantly online.

Let's stop for a second and contemplate this escalated and amplified mating dance—the baring of your soul to people you don't know. This paradox is sometimes called the *stranger on the train syndrome*, because it is a proven reality that people can feel more comfortable disclosing personal information to someone that they may never meet again (a great boon to spies, counterintelligence agents, investigative journalists, emergency room doctors, priests, and con artists). Other factors come into play too—namely the cyber effects of online disinhibition and anonymity. We feel less at risk of being hurt by a partner who has not seen us in real life. And in an urgent wish to form a bond, we disclose the intimate details of our lives without much hesitation.

But is it smart to disclose so much?

No. And this has been proven again and again. Telling perfect strangers intimate details of your life is guaranteed to make you more vulnerable—to criticism, to charges of narcissism and self-involvement, and, most troublingly, to fraudsters and criminals who are (I'm sorry to report) lurking and lingering online, searching for new ways to take advantage of you. Here's another reason why it's not smart: Our human instincts, which are driving us toward self-disclosure in order to form bonds, work differently online. We feel compelled to overshare and confess, but revealing too much personal information with a potential love interest online doesn't help to predict compatibility, the way it might in the real world.

Here's one reason: Hyperpersonal communication online, as described by Walther, is a process by which participants eagerly seek commonality and harmony. The getting-to-know-you experience is thrown off-kilter. The two individuals—total strangers, really—seek similarities with each other rather than achieving a more secure bond that will allow for blunt honesty or clear-eyed perspective. This explains why online support groups can be so therapeutic and nurturing. Online, judgment is suspended and people come to the support group with altruistic intentions of harmony and helpfulness. And these are easier to maintain if relationships remain in cyberspace. Why? Because when we are in a real-world support group, we can't help but react to real-world cues—facial expressions, posture, dress, body language, smell, tone—that might cause us to become self-conscious and wary of the judgment of others. We hold back. But when we are online, free of face-to-face contact, we can feel less vulnerable and not "judged." This can be truly liberating.

But for individuals using online dating sites or apps, the same conditions can lead to trouble. Similarities are magnified, differences are not considered, and this distortion becomes self-perpetuating. In the 1990s,

Walther discovered that when people meet in a technology-mediated environment, if an initial impression of someone is positive, then the subsequent blanks in the communication—unknown variables, lack of information, or separations—are filled in automatically with suitably positive and even totally idealized information. I would argue this isn't dissimilar from the way the human eye fills in gaps in the visual field to compensate for our natural blind spots. Connecting the dots in this way is called the *Gestalt principle of good continuation*. Online, you do the same thing as you build an impression of someone. Mentally, you begin filling in details about their personality, looks, and character to make a stranger more three-dimensional—and knowable.

If you're familiar with Freud, or psychoanalysis, then you probably know about the psychological effects of transference and projection—which cause us to unconsciously imbue others with traits or characteristics of people we've previously known—and you know how powerful this "filling in the blanks" process can be. (Hence the notorious exclamation, "You sound just like my mother!") The human imagination works with memories and impressions of people we've known in our past, which explains why the characters of so many great novels are usually based in part on real people the author knows or once knew. Transference is even more amplified in cyberspace.

Does this mean that the person you meet online is essentially a fiction, a partial fantasy of your own creation? I'm afraid so. Now add in the fact that on a dating site, it isn't even two "real" selves who are trying to meet and mate, but two "virtual" selves, two cyber artifacts, self-consciously constructed and curated for a particular effect. Authenticity is sorely missing. Princess Diana once said of her marriage that it was "a bit crowded" because there were three people in it. Even the standard experience of dating online involves four selves—two real-world selves and two cyber ones.

As two people online try to establish ties, they use their real-world instincts to determine whether they can trust each other or not. This sets them up for a real-world letdown and even a crash in self-esteem. (It isn't fun to suddenly sense that the instincts you use every day, in every way, are way off base.) Far worse, relying on their real-world instincts can lead vulnerable individuals into true danger. A woman who meets a man in a bar might never consider accepting a ride with him after only one encounter. Yet that same woman, after a few days of interacting through email and texts with a man she's met on an online dating site, may give out her home address because she feels such a strong connection with him.

This may partially explain a recent report, in February 2016, by the U.K. National Crime Agency (NCA), of a sixfold increase in online-dating-related rape offenses over the previous five years. A team that analyzed the findings presented potential explanations, including that people feel disinhibited online and engage in conversations that quickly become sexual in nature, which can lead to "misdirected expectations" on the first date. Seventy-one percent of these rapes took place on the first date and in either the victim's or offender's residence. What alarmed me most in this report was the apparent mutation of behavior. The perpetrators of these online date-rape crimes did not seem to fit the usual profile of a sex offender; that is, a person with a criminal history or previous conviction. This means we don't fully understand the complexity of online dating and associated sexual assault. But the cyber effects of syndication and disinhibition are clearly important factors.

The NCA does offer some helpful advice for online dating:

- Meet in public, stay in public.
- Get to know the person, not the profile.
- Not going well? Make excuses and leave.
- If you are sexually assaulted, get help immediately.

With all such risks and dangers, how do we explain the surge in popularity of online dating sites and apps? One of the drivers of online dating has to do with results: People who meet online are quicker and more likely to marry. When it's right, it's right. And with such a large number of people participating in online dating, the odds of finding a suitable mate only increase. In a survey of more than nineteen thousand people who married between 2005 and 2012, those who met on the Internet described their relationship as happier and more stable than couples who began their romance offline.

Another sign of success: The industry was profitable almost immediately. By 2007, online dating was bringing in $500 million annually in the United States, and that figure had risen to $2.2 billion by 2015, when Match.com turned twenty years old. By then, the website claimed to have helped create 517,000 relationships and 92,000 marriages and 1 million babies.

The sense of instant trust, or "instamacy," that comes with meeting in the cyber environment will be likely heightened in the future, when virtual-dating technology includes use of a full-sensory VR headset, DNA screening, and other subtle but significant information to make the matches. "Smart" glasses or contact lenses may track what types of individuals the wearer finds particularly attractive and factor this in to the matching technology. By all accounts, this is around the corner—perhaps a decade away. The future, in other words, looks heavy on science but a little light on love.

Swiping for Success

Nothing in recent years has proven the power of appearance and the first-impression bias more than the success of the dating app Tinder, which puts the two more important mating selection factors together quickly and brilliantly—proximity and attractiveness. The location-based service is in many ways a productive advance in online dating—a virtual cut to the chase.

On Tinder, you respond to profile photographs prior to any information. And you adjust your settings to find prospects in a proximal range that makes sense for you. Based on your location, photographs of prospects are provided. If you like a picture you see, you swipe right to learn more.

By 2016, a little more than three years after its launch, Tinder claimed to have generated 9 billion matches—more than the human population on earth—which suggests that either the whole world is using the dating app or some people are just really, really active. The stats are impressive: 196 countries, 1.4 billion swipes per day, 26 million matches per day. On the homepage, a photo collage of super-handsome young people fills the screen, with testimonials about how they owe their relationships and marriages to Tinder. *He told me I was a girl he would drive two and a half hours to meet. . . . He proposed after a helicopter ride. . . . Without Tinder, there would be no chance of us meeting.*

The process of swiping right for approval—and learning that your own image has been swiped right by someone else—has been described as "addictive" and even rewarding on a neurological level. But the promise of Tinder is that the app will fulfill your wishes for true love, and for adventure. The homepage sells romance as transformative: "The people we meet change our lives. A friend, a date, a romance, or even a chance encounter can change someone's life forever. Tinder empowers users around the world to create new connections that otherwise might never have been possible. We build products that bring people together."

Tinder CEO Sean Rad and I both spoke at a Web summit in Dublin in 2015. The venue was packed for Rad's talk, with thousands of people just waiting to be entertained by "hooking up" anecdotes, but his delivery was strangely flat, especially given the potential of the material. He gave a standard product pitch rather than regaling the auditorium with breathtaking insights into technology-facilitated mating rituals. And when he did wander into the subject of courtship, his advice was direct but not that profound. "The best piece of advice I give our users is be yourself," he said. "We underestimate our ability to look at a photo and pick up the nuances. When you're not yourself, people can sniff that."

Are we moving toward an era when technology-mediated relationships will be ephemeral—lasting only as long as a swipe? If swiping is the act that people find neurologically rewarding, it may be that they enjoy that more than actually finding a mate, or love. Some surveys suggest that face-to-face encounters between individuals—romantic or otherwise—are steadily on the decline. And there is evidence that dating apps and sexting, as well as other virtual encounters, actually may have a negative impact on people's sex lives.

When you fall for a face on Tinder, what are you falling for? As Rad suggests, do people really pick up on nuances, or what a cyberpsychologist would call "behavioral cues embedded in the image"? Psychoanalyst Dr. Robert Akeret has written books on the subject of how to analyze images and learn to see deeply into a still picture. In *Photolanguage: How Photos Reveal the Fascinating Stories of Our Lives and Relationships,* he interprets "hidden desires and fears, unspoken loves and hates, people's valiant efforts to camouflage themselves in the public world, and the inevitable leakage of their hopes and dreams and difficulties."

Leakage, now that's an idea. Never mind the language of love; what about the leakage?

Squinch!

As the success of Tinder proves, besides proximity, physical appearance is the most important driver in choosing a date and mate. In romantic relationships, appearance plays an especially critical role in whether a bond is formed or not. Obviously, this is subjective and personal. Beauty is (truly) in the eye of the beholder. But in study after study it has been shown that people tend to pair off with others of similar attractiveness, which means that an attractive person is also likely to draw a similarly attractive mate.

When stacked up against other traits like intelligence, warmth, or humor in the real face-to-face dating world, beauty almost always beats them out.

This doesn't mean that humans are supershallow and care only about looks. Attractiveness is a powerful asset in the competition for mates partly due to the *halo effect*. If a person is judged as attractive, it bleeds over and positively influences other impressions that are formed about him or her. This is why attractive people are judged as happier, warmer, and kinder, as well as more sociable, likable, successful, and intelligent. Attractiveness is a factor in almost every aspect of life and has been proven to give an individual huge advantages.

Here's how the halo effect works. Say, for instance, that you hear about a man you've never met who has rescued an old woman's cat from a tree. Would a bad guy rescue a granny's helpless animal? Surely not. This information puts everything else about the man in a positive light. This goes for certain professions, like emergency medical workers, firemen, and other first-responder jobs. It is hard to imagine selfish, insensitive people in these lines of work. Is it rational? No. The rules of probability demand there are all kinds of less-than-wonderful people in positive-halo-effect jobs, but being in these professions can cast a positive glow nonetheless.

Attractiveness works exactly the same way. Are there selfish, dimwitted, unsuccessful attractive people? Of course there are. But irrationally, human beings associate attractiveness as an overall positive sign. This could be because appearance is easy to judge. It is literally the first and most noticeable trait about anybody you meet, unlike the less obvious traits like honesty, integrity, reliability, and loyalty, which take some time to ascertain.

Making a quick judgment that someone is attractive is simple and, in terms of evolution, effective. Attractiveness usually indicates that an individual has fairly good odds of being healthy and fit, inside and out. The sheer convenience alone of this judgment call may be why it's so important.

How does this play out as a cyber effect? It is human nature to want to compete for dates and mates, to groom and try to improve our appearance and make a good impression, particularly when we are in dating mode. This is evolutionary reality.

As I described in the previous chapter about teens, the norms of cyberspace put even more emphasis on appearance—and how it is presented on social platforms. There is more pressure from the group to invest in appearance and self-presentation. Tinder's secret ranking algorithm—although they call it a "desirability score"—is a form of feedback loop. While sophisticated and exotic, it is basically ranking your cyber self—that carefully and consciously curated, filtered, Photoshopped, and otherwise enhanced self that many people now use online.

Almost anyone can achieve supermodel levels of beauty in cyberspace, particularly in a photo that captures only one quarter of the face. And in that regard, technology has been the ultimate equalizer in the genetic lottery of good looks. The norm online is now to see people who have been filtered—made thinner, taller, tanner, and blemish-free. This means the techno-evolutionary competition to appear attractive has grown even more intense. Survival of the fittest now requires expert Photoshopping skills. There is little doubt this will continue.

As the cyber norm becomes more attractive, the halo effect of attractiveness is escalated online as well, partly because of the importance of first impressions and some of the laws that govern them. One of these is the *primacy effect,* which describes how one attractive trait or feature of a person you meet will jump out and grab your attention—and drown out all others. "Oh my goodness, his eyes are so blue!" "Wow, she has lips like Angelina Jolie!" When you consider your own appearance, what is your best feature? That's your primacy effect. If you are going online to date, you should know it.

In cyberpsychology, two constructs are used to describe the power of impressions made online. When you make assumptions about a person based on their Match.com profile photo, for instance, that is known as *impression formation.* And when you filter and fix and curate your own profile photo on Match.com, it's *impression management.* The mere act of choosing which picture to use on a dating site—active, smiling, unblemished, or nostalgic—requires that you imagine how you look to others and aim to enhance that impression.

Like it or not, there are arguments for doing just that—if a real-world meeting is not likely. Because if people find your profile photograph attractive, they will form a better impression of you, and that impression can have a strong and lasting effect on the other judgments they will subsequently make about you.

But an online dating site is different. Individuals need to communicate attractiveness but not alter their profile photographs to the point of rendering them unrecognizable in real life. At the same time, a profile photograph on a dating site or app creates an impression—made in roughly forty-two milliseconds—and that impression will cause a bias.

Science being science, and considering how obsessed with appearance people are on dating sites, there has been good research done about what kind of profile photographs lead to the best first impressions. This is the intersection of science and digital media, and as serious as that sounds, I also find it pretty funny. There is no harm—and actually some good—in passing along the findings based on the best research available.

IMPRESSION-MANAGEMENT TIPS FOR YOUR PROFILE PHOTO

- Wear a dark color.
- Post a head-to-waist shot.
- Make sure the jawline has a shadow (but no shadow on hair or eyes).
- Don't obstruct the eyes (no sunglasses).
- Don't be overtly sexy.
- Smile and show your teeth (but please no laughing).
- Squinch.

Wait.

What's a squinch?

It is a slight squeezing of the lower lids of the eyes, kind of like Clint Eastwood makes in his Dirty Harry movies, just before he says, "Go ahead. Make my day." It's less than a squint, not enough to cause your eyes to close or your crow's-feet to take over your face. If you want a tutorial on how to produce the perfect one, then I can recommend one by professional photographer Peter Hurley, available on YouTube, called "It's All About the Squinch."

Trolling, Catfishing + Cybercharming

It probably goes without saying that online dating has been around long enough for some individuals to have become quite expert at making an inordinately good impression in cyberspace. They have their squinch down, and can even charm you, initially. These are deceptive skills that can work for them in real life too. While some individuals may use cyber-dating to experiment with new selves, new behaviors, or a new gender, there are other people who just like to lie about who they are—and trick strangers. It isn't an "honest mistake" or misunderstanding arising from experimentation. It is simply fraud and trickery.

A famous case of what is known in cyberpsychology as *identity deception* occurred early in the rise of the Internet, on a women's discussion board. One participant in the forum, Joan, described herself as wheelchair-bound and mute—which explained why she could not attend face-to-face meetings. Online, she expressed herself and her desires to find romance—and earned the trust of other women on the forum, some of whom were also looking for a relationship. Helpfully, Joan introduced these women to an able-bodied friend of hers, a psychotherapist named Alex, who won the trust of the women in the forum and, eventually, had sex with them. Unfortunately, though, it was discovered that Alex was actually Joan too. He had invented the Joan persona online, he claimed, to better understand his patients. But he took the role too far—and just couldn't stop being Joan.

Another kind of trickery online is called *catfishing,* which has been dramatically illustrated by an MTV series in which each episode follows, reality TV–style, two people who are having a relationship solely online—sometimes using a false name, false age, or false gender. The term was coined in 2010 with the release of a compelling documentary, *Catfish,* about Nev Schulman (now the host of the MTV show), who fell in love with a beautiful and vivacious girl in a small Midwestern town whom he "friended" on Facebook. They corresponded for months. When he went on a road trip with a documentary film crew to meet her, he discovered that she wasn't real. Instead, she was the creation of an unhappy and seriously overweight forty-year-old mother of three. She felt just terrible, she claimed. And wasn't out to hurt anybody. But, of course, she did hurt somebody.

Yes, there have always been tricksters, con artists, and liars among us who pretend to be somebody they aren't. But technology has made it so much easier. You can set up new email addresses in minutes, conjure letterheads, business cards, an email string, and contact list, and make hundreds of Facebook friends in a day. You can create a convincing profile and social-network presence harvested from Google image searches or Facebook. (This is all made possible because content is difficult to secure.) Once upon a time, deception required skill and, I hate to say it, even artistry. Now you don't need the imagination of a Tolstoy or Dickens to create a totally believable but fictional identity. It's a matter of cut-and-paste.

Another type of trickster online is more malicious. As discussed in previous chapters, *trolls* are individuals who deceive for the joy of deceiving—stimulating controversy and taking people by surprise. For those who use dating sites, trolling can be a real problem. Trolls on dating sites are capable of anything from sexual harassment, lewd comments, and name-calling to cyber-exhibitionism and threats. The motivation can be similar to catfishing—often a combination of boredom, attention-seeking, revenge, pleasure, and a desire to cause disruption and damage. It is clearly aggressive and sometimes sociopathic. Women are often the victims.

A smart solution to catfishing has been created by the dating app Blume, launched in 2016, which requires users to put up a "real-time selfie" as a way to authenticate identity.

Another new dating app, Bumble, seeks to solve the online dating troll problem. It was founded by Whitney Wolfe, a cofounder of Tinder who left the company and filed her own sexual harassment lawsuit against it. Bumble uses profile swiping to match single people in the same town, similar to the way Tinder works, but in this case women are forced to make the first move. Within twenty-four hours after being messaged about a match, women must respond or the match disappears. Wolfe felt strongly that women were disadvantaged on dating sites and decided to empower them, inspired by Sadie Hawkins dances.

"We're definitely not trying to be sexist, that's not the goal," Wolfe says. "I know guys get sick of making the first move all the time. Why does a girl feel like she should sit and wait around? Why is there this standard that, as a woman, you can get your dream job but you can't talk to a guy first? Let's make dating feel more modern."

Any time a large and vulnerable population can be easily targeted, there will be those who seek to take advantage of them, so I can't say that I'm surprised by the existence of fraudulent identities online, as much as I'm deeply troubled by the cavalier attitude about them. The lighthearted manner in which trolling is discussed exposes a corrosive cynicism and lack of regard for honesty. The approval, even cheering, for acts of harm and deceit against real and innocent people (this is not a movie, folks) only shows how quickly abnormal, unkind behavior can become normalized in an environment if enough people look the other way, and even applaud. The American literary critic Norman Holland identified this phenomenon as early as 1996 and named it *Internet regression.*

Among online enthusiasts, and in some tech cultures, it's considered really cool to be a troll. A similar sensibility led to the escalation of aggressive cyberbullying. Since when has hurting people become a sport, or fun? Of all the many evolving norms, this is truly one to worry about. Not just the feelings of the victim but the psychology of the perpetrator.

A Canadian team of psychologists did an interesting study of individuals who comment frequently online, asking them to complete a personality inventory and a survey of their commenting styles. The results are disturbing and show strong positive associations between individuals who comment frequently, enjoy trolling, and identify themselves as "trolls." There was also a relationship between those who liked to troll and three of the four components of what is known as the *dark tetrad of personality,* a set of characteristics that are found together in a cluster: narcissism, sadism, psychopathy, and Machiavellianism. The researchers described trolling as a manifestation of "everyday sadism." Now, that is truly something to avoid online.

Lastly, I need to discuss a seemingly more benign personality type to watch out for online. The word *narcissist* is commonly used—and thrown around—these days, partly because there seems to be more awareness and discussion of narcissism in general. Like other aspects of personality, there is a spectrum of behavior. But there is also a diagnosable condition called *narcissistic personality disorder,* which is a good thing to know about if you are dating online. Here's the Mayo Clinic description: conceited, boastful, pretentious, monopolize conversations, perceive others as inferior and have a sense of entitlement. They become impatient and angry when they don't get special treatment. They can't handle anything that is perceived as criticism. . . . Causes may be excessive parental criticism or praise. Also genetics and psychobiology. . . . Nature or nurture or both. More males than females. Often begins in teen years.

Narcissists are preoccupied with fantasies of success and romantic notions about finding the perfect mate. This is not because they want to find true love. What they want is a trapping or trophy that will enhance their feelings of self-importance and make them look better. To achieve this, a lot of searching and dating and a good deal of swiping are required. And due to the fact that narcissists have trouble remaining in relationships, together with their adeptness at manipulating social media, there is likely a greater percentage of them appearing on dating sites than you'll find in the offline dating population.

Another reason: The very qualities that make a narcissist undesirable in real life—selfishness, arrogance, self-centeredness, and a need to feel better than you in every way—can actually give them a leg up online. They are now more competitive in the mating selection process, because the established cyber norm is to parade your accomplishments, post pictures of your expensive car and your gorgeous house, and document every meal of your fantastic travels around the globe for all your cyber-friends to see. Most important of all: posting lots and lots and lots of pictures of yourself looking good.

Narcissists need admiration, flattery, and loads of attention. They need an audience. This is why cyberspace is made for them. Their highest highs can come from making people fall in love with them. The problem is, given the way they ooze confidence and cybercharm, it may be harder to spot them—and know to stay away. To help you out, here's a mini-inventory of questions to ask yourself:

- Do they always look amazing in their photos?
- Are they in almost all of their photos?
- Are they in the center of their group photos?
- Do they post or change their profile photo constantly?
- When they post an update, is it always about themselves?

Cyber-Infidelity

Having an adulterous affair used to require fluke, luck, and a lot of hiding in shadows. People had to hope for chance meetings or fabricate work conferences and weekend retreats to cover their tracks—then check into strange hotels under false names, pay cash, and sneak out before sunlight. Not everyone is cut out for such subterfuge.

But for individuals who struggle to stay in a monogamous relationship, technology has made the temptation much, much greater. You can reach out from the comfort of your own home—and flirt on a bad hair day and in your bathrobe. In an hour, you can create a new profile on a site, create a new email address and a pseudonym, and have a date in front of your webcam. Nobody will know but you—and your cyber-fling.

There has been a societal change in cyber-dating for those who are married who desire a discreet affair on the side. There are now a growing number of these sites and apps in operation—homepages with FAQs such as *Why have an extramarital affair?* that are answered with seductive rationales. The normally unspoken rules of an affair are often part of the "terms and conditions" that must be agreed to before becoming a member of the site. Discussed so blandly, and practically, it is another way of normalizing or justifying behavior that is considered by most people to be unethical and deceptive.

(On a side note, it's worth knowing that you aren't the only one who never reads these terms-and-conditions agreements carefully. An experiment was done in a London café with a free Wi-Fi hotspot in 2014. Six users clicked the box to accept the terms and conditions even though, by doing so, they were agreeing to "assign their first born child to us for the duration of eternity." Needless to say, the term was not enforced.)

For some, having a cyber-affair is a way to practice and "pretend" infidelity. You may have fantasies about straying from your marriage, but by keeping it online, you can feel safely distanced from the actual behavior and deception. *Hey, it wasn't me at all!* The online self is compartmentalized, a type of *dissociative anonymity*. It's a convenient way to stay guilt-free.

It is well known in military psychology that killing from a distance is considered easier than killing at close range. This explains why it is easier for most people to push a button that sends a droid or smart bomb that is targeted to hit people far away than to face another individual and witness the effects of your aggression. The greater the distance you are from the people you are harming, the easier it is on your conscience.

That's why I suspect that, in cyber-affairs, the perceived distance between the lovers helps them to depersonalize and not consider the people they are hurting.

When cyber-infidelity ceases to be enough—and there is an escalation effect involved in infidelity, like all deceits (the slippery slope)—you can move it into real life without missing a beat. Now with apps like Hinge, Lavalife, Grindr, meet2cheat, Affairs Club, and Ashley Madison, which were all designed specifically for casual hookups and affairs, you can easily syndicate online to find cohorts to be unfaithful with. But cyber-cheating doesn't come without risk. The hacking of Ashley Madison in 2015 was disastrous for its 30 million adulterers, but at the same time, it provided a treasure trove of incredible data for researchers. Annalee Newitz of Gizmodo reported a massive gender divide when it comes to extramarital affairs, pointing out that while 20 million men had checked messages on the site, only 1,492 women had. Ashley Madison has since denied these figures, arguing that many more women used its site. In an interesting cyber twist, it appears that extensive use of female chatbots may have confounded interpretation of the data.

Geographically and culturally, there are other interesting data for analysis. "Italian users were most likely to be looking for a short term relationship," according to *Wired* magazine, "while German and Austrian users preferred a long term arrangement. Chinese users most frequently opted to keep things online, expressing a preference for a 'cyber-affair,' something of no interest to Japanese and South Korean users."

Untold numbers of divorces and breakups resulted from the hack, but there were much larger consequences. Given the release of private information into the public domain, thousands of individuals will now be blackmailed, and two people have already committed suicide. And there were eight thousand Ashley Madison users in the country of Pakistan, where adultery is illegal. People who signed up must have recognized that there is always risk associated with adultery, but something seduced them into participating anyway. Perceived anonymity is a powerful cyber effect. And risk, as we know, cyber-migrates from online to real life. Unfortunately in Pakistan, that could mean users will be migrating to jail. As one meme suggests, Ashley Madison should change its motto from "Life is short. Have an affair" to "Life is short. Hire an attorney."

Sex can be like a drug. So can romance. Technology has re-created the same conditions—availability and constant temptation—that plague (or delight) movie stars, rock singers, and professional athletes, who may try to be monogamous and faithful but aren't superhuman after all. For an individual who struggles with impulsivity or compulsive behavior, the promise of these easy-affair apps can create serious problems.

As early as 2009, *The Guardian* newspaper reported that a Facebook affair or cyber-infidelity was the listed cause for a growing number of divorces in family law practices in the U.K. Several stories hit the media that year—a woman who discovered that her husband was having virtual sex with another man, a woman who discovered her husband was having a Facebook affair, and a British couple who were divorcing after the husband was found cheating on his wife with a character on *Second Life*.

The ethics of cyber-infidelity is a burgeoning area of the law, which is evolving differently, country by country. In the U.K. and Canada, sexting, lurid Facebook posts, and other manifestations of infidelity do not qualify as "adultery" per se—because sexual intercourse has not taken place in the real world—although these may serve as a reason to end a marriage. In the United States, excessive time spent online, whether it's for cyber romance or gaming, has been used in court to show a history of poor parenting and has impacted the amount of child custody granted.

You can't call it adultery, but that doesn't mean that cyber-infidelity isn't a serious breach of trust. These indiscretions have consequences that are very real. People who have wandered away from a committed relationship and family life in order to seek love online or even just virtual sex can leave their partners and families feeling abandoned and victimized. A special issue of *The Journal of Treatment and Prevention* discussed the emotional outcomes of virtual distractions and disloyalties in a compelling article called "Is It Really Cheating?" which convinced me of the resulting trauma of this sort of betrayal.

Since real-world emotions play out online, where there's cyber-infidelity, there's also jealousy—and "self-help" websites informing spouses of how to look for telltale signs of a partner cheating online.

Multiple studies have shown that social-network activity does trigger and amplify jealousy among users, which brings me to a quick murder story. It has to do with the uncontrollable cyber-jealousy of Giuseppe Castro, a man in Italy who decapitated his wife when he suspected she was having an online affair.

"She was always chatting with other men," he told police. "I couldn't take it anymore."

Virtual Girlfriends

LovePlus and its sequels are Japanese video games that offer avatar love for a handheld Nintendo DS or for Apple's iOS. These games are marketed as "dating simulators" that help individuals learn to be in a relationship, and possibly even learn to love. Since its launch in 2010, *LovePlus* has attracted hundreds of thousands of players worldwide, from middle-school-age boys and girls to adults who prefer the company of one of the three sweet unreal girls offered by the game—Rinko, Manaka, or Nene—rather than deal with the headaches of being with an actual human being.

Rinko, Manaka, and Nene are adorable and devoted. And their personalities can be adjusted in settings. Best of all, unlike the cyber-melodrama *Her,* they are programmed never to dump you. It's just not in their code.

Players have declared that their love for these virtual girlfriends feels real—and is consoling, supportive, and nurturing. We know from studies that people can form real and authentic emotional attachments to virtual characters. An avatar will always reflect an aspect of its creator and could be, in some ways, more revealing than a self-consciously manufactured and curated self-portrait.

Some have used the video game as a refuge when going through a bad breakup. "I would say that a relationship with a *LovePlus* character is a real relationship," said anthropologist Patrick Galbraith, who specializes in Japanese popular culture. "People are really intimately involved."

But can we really call it love when there's nothing to lose? Would anybody ever write a song to Rinko, Manaka, or Nene? If human beings become used to unconditional machine love, what does it mean for their emotional resilience after a real breakup?

Let's take this a step further. If human beings begin to prefer demure and devoted Japanese anime characters over real women, what does this mean for the future of the human race? As Norman Holland writes, this parallels the sci-fi story *The Stepford Wives,* a male fantasy about women who are completely satisfying because they are docile and have no real needs of their own because they are machines.

In order for our species to survive and thrive, people need to mate with other people, not just fantasize about Rinko, Manaka, and Nene. In Japan, this concern is no longer mere speculation. A government survey released recently estimated that nearly 40 percent of Japanese men and women in their twenties and thirties are single, not actively in a relationship, and not really interested in finding a romantic partner either. Relationships were frequently described as "bothersome."

Another survey found that one in four unmarried Japanese men in their thirties were virgins. The number of virginal single women in their thirties was only slightly less. At the time, Shingo Sakatsume, who works as a "sex helper" and counsels middle-aged virgins, observed, "In Japanese society, we have so much entertainment beyond love and sex. We have animation, celebrities, comics, game, and sports. . . . Why do you need to choose love or sex over the other fun things that don't have the potential for pain and suffering? The illusion of a perfect relationship, combined with the Japanese fear of failure, has created a serious social problem."

By 2060, if current trends continue, Japan's population will have shrunk by more than 30 percent.

Cyber-Celibacy

Don't assume the explosion in virgins is restricted to Japan. There are a number of factors online—invincibility for starters—that could create a heightened fear of failure in the real world and a wish to avoid pain by restricting human contact to virtual relationships. And the less practice the population gets with face-to-face interactions, presumably the worse they will become at them—only amplifying their vulnerability. Cultural considerations aside, what is happening in Japan could come to pass everywhere.

Intimacy comes with a risk and can be difficult for some. For those who have had painful experiences in early relationships, or difficulty during a developmental stage, it can feel too dangerous, too risky. In her book *Alone Together,* Sherry Turkle provides some insight as to the appeal of tech and its ability to arrest or mitigate fears of intimacy. "These days, insecure in our relationships and anxious about intimacy, we look to technology for ways to be in relationships and protect ourselves from them at the same time."

I wonder, though, if technology itself could be keeping us from learning how to be intimate—and feel comfortable in a close, caring, face-to-face relationship. Some surveys suggest that face-to-face encounters between individuals—romantic or otherwise—are steadily on the decline. And there is evidence that dating apps and sexting, as well as other virtual encounters, actually may have a negative effect on people's sex lives—and, importantly, what we traditionally call human intimacy. Dr. Craig Malkin, clinical psychologist at Harvard Medical School, makes a case that technology itself can make us afraid of intimacy:

> Strangely enough, our brains don't seem to care if the thrill comes from great sex, drugs, or an epic win in World of Warcraft; they all cause massive amounts of dopamine, a feel-good neurotransmitter, to start spilling into our brain's reward center. And being something of a neurological prime mover, dopamine tends to keep us chasing after the same thrill again and again, regardless of the consequences. So while gaming or pornography can't ever cure our loneliness, over time they do become an incredibly addictive salve—and *that* makes it easier and easier to turn away from people and back to cyberspace.
>
> The end result is that much as people in pain sometimes drown their sorrows in alcohol, the cybercelibate abuse technology, relying on it to provide relief, relaxation, self-soothing, excitement, and even connection (albeit limited) that they could be getting from live people. . . .

> And that's why cybercelibacy is a problem for all of us. We're all a *little* anxious about intimacy, aren't we? Some researchers have even suggested that technological isolation is at least one contributing factor to the decline in marriage and committed long-term relationships.

In other words, despite everything I said in chapter 1, and all the evidence that we might be in the midst of a sexual revolution, the current era of the easy hookup appears to have created more intimacy online, and perhaps even more sexual activity, but less actual contact sex. As Turkle puts it, more people are "alone, together." In an age of technology, this is the new definition of sexual intimacy.

The Uncanny Valley

Romantic love has never been easy. I guess you can't blame people for hunting down some simpler solutions, or for technology to keep designing them. The search for something that approximates real romantic love and companionship (but without the problems of dealing with actual people) leads technologists and the industry to create artificial-intelligence solutions—robots programmed for all kinds of purposes, to provide anything from comfort to therapy.

The field of cyber-love and robotics is growing, and the sexbot industry is now attracting big money, tech innovations, and high-level artistry—to make the bots as real as real can be. In fact, they are so real that one Japanese manufacturer of the humanoid robot, a cute puppy-eyed social companion called Pepper, had to take the unusual step of creating an urgent "term of condition" for users: "The policy owner must not perform any sexual act or other indecent behavior" on the machine, which is designed to live with humans. One dissatisfied owner, after an alcohol-fueled rage, was arrested after assaulting Pepper.

As for love and companionship—a real relationship—someday, just like in the movies, there may be avatars and bots that mimic humanness so well that real human beings fall head over heels in love with them. For now, though, there's a well-known tech-design problem getting in the way. In the film industry, this was first discovered and documented in 1988, when Pixar screened a computer-animated short film called *Tin Toy* and learned that the test audiences hated the sight of a realistic human baby who terrorized the toys. The reaction was so strong that Pixar decided that, going forward, it would create animations that weren't too closely human in appearance. Why would an animation that is "too close" to human repel audiences?

"We still don't understand why it occurs or whether you can get used to it, and people don't necessarily agree it exists," said Ayse Saygin, a cognitive scientist at the University of California, San Diego, to *Scientific American*. "This is one of those cases where we're at the very beginning of understanding it."

Some argue that a perfect replica might be acceptable, but an approximation is just too creepy. "Pixar took a lesson from *Tin Toy,*" said Thalia Wheatley, a psychologist at Dartmouth College. "We have to nail the human form or not even go there."

Filmmaking, especially animation, relies increasingly on the technology of computer-generated imagery, or CGI, which is capable of producing a close facsimile of a human being. But those avatars in the film *Avatar* were blue for a reason—and audiences found them beguilingly beautiful. And audiences and reviewers reacted negatively to the train conductor and other characters in the animated film *The Polar Express.* (Was that Tom Hanks or not?) Rather than replicating humanness, these facsimiles seem to leave people feeling queasy or sick, a reaction not unlike one that some people experience when they see bad plastic surgery.

Rather than admiration, many people respond to too-close human replicas with repulsion.

This phenomenon even has a name, the *Uncanny Valley.* The concept was first introduced in 1970 by Masahiro Mori, robotics professor at the Tokyo Institute of Technology.

Mori, now in his late eighties, became involved in robotics some fifty years ago, before the term *robot* or *robotics* was even applied to his field. He designed one of the first three-fingered artificial hands that was used as an industrial tool for handling radioactive materials and later was developed as a prosthetic aid for disabled people.

Working on the robotic hand, he first experienced what he described as an "eerie sensation." To scientifically illustrate the emotional effect of seeing the hand, he plotted a graph with "affinity" on one axis and "human likeness" on the other, to map how people reacted to various robotic prototypes. The chart showed a low human affinity score for crude industrial robots (which look nothing like humans) and a medium affinity score for toy robots (who can look quite cute in a mechanical way, like R2-D2 or C-3PO in *Star Wars*). But when robots become too humanlike, affinity crashed, causing a giant dip on Mori's graph. Hence the name, the Uncanny Valley.

In case you are wondering if this revulsion is culturally conditioned, the answer is no. Even monkeys appear to have an Uncanny Valley reflex: An experiment found them looking longer at real monkey faces and unrealistic monkey faces than they looked at realistic synthetic faces. This means it is primal—and hardwired. But why?

Mori's observation has become widely known and discussed in tech circles only in the past few years, but his insight is timeless. "I think this descent explains the secret lying deep beneath the Uncanny Valley," he wrote. "Why were we equipped with this eerie sensation? Is it essential for human beings? I have not yet considered these questions deeply, but I have no doubt it is an integral part of our instinct for self-preservation."

And he speculates on the cause: "Proximal sources of danger are corpses, members of different species, and other entities we can closely approach."

What Mori brings to science and research is unusual. He is listening to his own intuition and the mysterious but documented responses of many others. Without a doubt, he assumes there's a reason for this pervasive feeling. In the discipline of science, the opposite is expected. We aren't supposed to assume anything until there are studies and research evidence to back it up.

But Mori is saying something else. He is saying: Pay attention to the human feeling. It is there for a reason. And he recommends that we pursue nonhuman designs, and respect the intuitive yet mysterious human reaction to fakeness. Otherwise, why would we be equipped, as he puts it, with this uncanny sensation?

Isn't it possible that there are essentials for humans that we need to pay attention to before we have conclusive studies? This is something I am passionate about, and a perspective I feel is needed at a time when technology will surely keep outpacing science. In the absence of studies, we need more common sense.

And respect for the human.

Some in the robotics and CGI industries view the Uncanny Valley as a design challenge, and are determined to solve it. What happens when they do? Imagine what the future could hold, when humans are emotionally targeted by the manufacturers of artificial-intelligence machines. Like the games and dating apps we've grown to use and love, and depend on for entertainment, pleasure, and intimacy, the robots will likely be seductive and irresistible. It could start almost benignly, with an adorable robot or avatar that you can't help but love. Artificial intelligence, whether it is embedded in a robot or in a Tinder algorithm, can impact human life on the most profound levels—from finding a mate to intimacy. We are moving from natural selection to cyber selection.

When it comes to technology, should we be trying to design beyond human instinct just because we can? I suspect that the Uncanny Valley is an evolutionary red flag—a warning call. Human beings have an innate intuitive fear of snakes and spiders, of heights and too-tight spaces (as well as corpses, as Mori points out). These aversions aren't pointless. They are primal fear—instinctive reactions for self-preservation and species propagation. These aversions evolved to protect us, and they have served us well, from contact with prehistoric Neanderthals to our complex relationships with fellow *Homo sapiens*. Perhaps we need to pay more attention to what Mori warns us about—the little things. And a much bigger thing too. Love.

CHAPTER 7

Cyberchondria and the Worried Well

The imagination is a beautiful thing. Scientists believe it has an evolutionary purpose, that we acquired imagination to help us anticipate danger. As our prehistoric ancestors struggled to survive in a threatening world, they learned to envision potentially dangerous situations—*If I walk past that clump of trees up ahead, I might be attacked by a beast hiding inside it*—and how best to avoid them. By imagining a worst-case scenario, we can control dread and anxiety and are able to make vital decisions with mental clarity and calm.

Something changes, though, when we go online to search for medical information. When we try to find out about a physical symptom we are experiencing—or worried about—what appears to be a useful exercise to help ease anxiety only escalates it. And as a growing body of evidence suggests, it may actually be making us sicker.

Today's medical doctors know this phenomenon well. More and more of the patients they see in their offices and clinics arrive consumed by fear that they are terribly ill. Clutching their "Google stack," as it's sometimes called, these individuals have been "researching" their symptoms and bring evidence to support an amateur medical opinion that's been formed by surfing the Internet. A small number of doctors actually welcome this collaborative approach. But by and large, this is not the case—because it can interfere with the proper diagnostic process. This is particularly true if there is no physical evidence. A patient can go online and compare his or her own skin rash with images in a catalog of dermatitis and may be able to come up with some helpful matches. But in the absence of physical symptoms, it is speculative and unhelpful, and this is also true regarding mental health conditions. It may be fascinating for some people to complete online psychometric evaluations regarding anxiety or problem drinking, or even find themselves doing a "desktop" self-diagnosis of schizophrenia or borderline personality disorder, but only qualified professionals should be diagnosing such conditions.

Many medical colleagues have described to me the great frustration they feel when patients claim to know the diagnosis prior to their examination. Often the first part of the consultation is spent convincing these people that their hypothesis is flawed: "No, you do not have dengue fever" or "infectious river blindness" or some other disease found only in an Amazonian jungle.

What should be good news is often met with only mild relief, even tinges of disappointment. Patients don't like to be told their super-sleuth medical detective skills haven't delivered. Meanwhile, out in the crowded waiting room, the clock is ticking and the truly sick are kept waiting for their consultation.

One reason that health coverage costs are so high: Reports state that up to $20 billion is spent annually in America on unnecessary medical visits. How many of these wasted visits are driven by a cyber effect?

There are benefits to being vigilant about health. And the Internet offers so much good science and medical information that it should be helping us stay healthier and happier than ever. Sadly, it often doesn't work out that way. Consider the story of Lisa—a healthy, thoughtful, well-educated woman not given to impulsive behavior—and her virtual descent into the world of Lyme disease.

Lyme Phobia

Lisa, a woman in her mid-forties, went to Cape Cod, Massachusetts, a few summers ago to see an old childhood friend, Michelle. The women were on the planning committee of their thirtieth high school reunion and used their time together to begin their work. After a morning of brainstorming and too much coffee, they needed to clear their heads, so they took a walk in the woods along the bay. Lisa had heard there was a lot of Lyme disease on Cape Cod, and asked Michelle if it was safe. It was high tick season, in fact. The blood-drinking parasites in the wooded areas are most prevalent during the spring and late summer.

Just the word *tick* made Lisa feel itchy, as if the bugs were crawling all over her. Indeed, that evening before bed, she found a small deer tick, the size of a sesame seed, attached to the back of her neck. Unsure how to pull it out—pinch it with tweezers or between two fingernails?—Lisa went online, trying to remain calm as she clicked from one search result to another. With the tick happily feeding, Lisa's online investigations took her to a public information website created by the U.S. Centers for Disease Control and Prevention. Not wanting to wake Michelle, who was asleep in another part of the house, Lisa followed the step-by-step instructions provided by the CDC, and with clinical focus and the help of her surgical instrument (tweezers) and two mirrors, she extracted the hapless deer tick and flushed it down the toilet. For her "post-op," she rubbed alcohol on the bite wound.

The tick was gone, but Lisa's anxiety was rising. Stressed-out and unable to sleep, she decided to read more about Lyme disease, a tick-borne illness that's been in the upper Midwest and Northeast regions of the United States since the late 1970s, when an improbably large group of Connecticut schoolchildren were diagnosed with juvenile arthritis, and it was traced to infected ticks found in the region.

One of the frustrating things about Lyme is the variety and murkiness of symptoms, and resulting diagnostic difficulties. Debates rage about the disease, in particular whether there is truly something called *chronic Lyme disease syndrome* and what the proper treatment should be. Lisa had heard somewhere—probably in one of the media stories about the disease she'd read—that if left untreated, Lyme could have dire consequences.

Click after click, as she tumbled deeper into the world of medical search, reading about the consequences of Lyme, from mood swings to meningitis, Lisa was doing what a lot of people around the world now commonly do when they become sick, or believe they might be sick. In a large international survey, a majority of people said they used the Internet to look for medicine, and almost half admitted to making self-diagnoses following a Web search. A follow-up survey the following year found that 83 percent of 13,373 respondents searched the Internet often for information and advice about health, medicine, or medical conditions. Individuals in "emerging economies" used online sources for this purpose the most frequently—China (94 percent), Thailand (93 percent), Saudi Arabia (91 percent), and India (90 percent) led the table of twelve countries.

Worrying about one's health—or that of a loved one—is perfectly normal. To some extent, it's a natural outgrowth of an awareness of one's own physical vulnerability and mortality. In cultures across the globe, when friends and family gather to lift glasses in a toast, what are they most likely to say? "To our health!" In Ethiopia, it's *Le'tenachin!* In Arabic, *Be sahtak!* In Bosnian, *Nazdravje!,* and in Czech, *Na zdravi!* In Ireland we say, "Health and long life to you, land without rent to you, a child every year to you, and death in Old Ireland." (That bit about "land without rent" is a throwback to our days as a colony. Just like the Internet, Ireland is a place that cannot forget—we've even found a way to politicize a popular toast to health.)

The essential question: When does a normal desire for health, and the willingness to pursue it by combing all available information, turn dangerous? When do the "worried well" literally worry themselves sick?

Unfortunately, this happens very quickly these days, often aided and abetted by technology.

Twenty years ago, if you were experiencing the onset of any physical condition that persisted to the point of interfering with your activities and work, you would visit a doctor's office and consult a bona fide expert with years of rigorous medical training, a reassuring manner, and a selection of impressive diplomas hanging on the waiting-room wall. In the digital age, we do our own symptom search, as Lisa did. It's a layman's game, like "playing doctor" at home. When you're feeling crummy, the first step is often to seek an answer online by consulting one of many medical websites available, many bearing seals of excellence and brand names that radiate credibility. It's a virtual magic box of medical marvels, ranging from the impressive online presentation of the Mayo Clinic to the opinionated offerings of personal blogs and chat forums. The choice is seemingly infinite.

Cyberpsychologists have learned that people make judgments online in a number of ways. About half of the medical information offered on the Internet has been found by experts to be inaccurate or disputed, but the criteria that online searchers use to determine their level of trust is pretty superficial: the overall design of the website, the brand recognition of the URL, and whether the advice appears to come from experts or those with shared experiences, rather than the integrity of the information itself. There are even design awards for "best health site" granted to the likes of WebMD.com with its 40 million unique monthly visitors, NIH.gov with 22 million unique monthly visitors, and MayoClinic.org with 17 million. With their pristine, clinical look—and enormous popularity—online medical websites seem like reliable sources of advice.

Lisa lost track of time—a few minutes became an hour, then another. In a state of heightened anxiety, her imagination ran wild and she ignored information that might have been comforting, like the fact that most ticks are not dangerous. Only small deer ticks carrying the bacterium *Borrelia burgdorferi* can transmit the disease to a human being, and only after being attached for thirty-six to forty-eight hours. Even though her walk in the woods had been only twelve hours earlier, and the tick likely had been attached for less than eight hours, Lisa continued to worry—and search for more information. Better safe than sorry!

She learned that the first sign of disease is commonly a red bull's-eye or target, called *erythema migrans,* located where the tick's mouth was attached to the skin. As Lyme progresses, the stricken individual can present a number of common symptoms within a week to a month—sore throat, stiff neck, fever, chills, rash, and severe body aches. It can feel like you're getting the flu. Just thinking about it, Lisa felt hot and flushed.

She put down her laptop, returned to her mirrors, and, by adopting a pose worthy of a contortionist, glimpsed some redness in the area of her tick bite. This only made her worry more. After all, if left untreated,

Lyme disease can lead to debilitating arthritic conditions, clinical depression, encephalitis, facial paralysis, acute inflammation of the tissue surrounding the heart, and other complications that, in some cases, although rare, can lead to death. In one five-year period, there were 114 reported fatalities resulting from complications of Lyme in the United States. This is fewer than half the number of people killed by lightning strikes over the same period, but Lisa didn't know that.

Which was worse: facial paralysis or death?

Lisa wasn't sure.

Without a doubt, having a parasitic insect attach to your flesh, burrow in, and drink your blood is unsettling in itself. But reading the medical websites and forums frequented by Lyme disease sufferers took Lisa to an even more frantic and disturbed place. Forum participants used a clubby insider-speak to tell their personal stories, talking offhandedly and even competitively about their advanced cases of Lyme and how many treatments they had tried in vain.

I'm sure all of us, at some unfortunate time in our lives, have sat in a doctor's waiting room in the presence of a person I call Catastrophia, the Mother of All Patients, who chats you up with a catalog of unrelenting horror that is her medical history: "And just when I thought it couldn't get any worse . . ."

Silently you scream *Enough!* but you are a captive audience, and social convention dictates that you sit there politely and muster a nod.

The point is, most people who've been ill do make a good recovery and get on with their lives. Yet there is a minority who like to linger there and dwell on illness—enjoying the attention they get from regaling others with their medical nightmare stories. They seek an environment where they can broadcast their pain and suffering. Technology now offers them a doctor's waiting room in the cloud. The online health forum, where an audience of thousands awaits, is tailor-made for the chronic complainer and, worse, the truly mentally ill.

The treatment of Lyme is actually quite easy and effective, if caught early enough—three to four weeks of the antibiotic doxycycline usually takes care of it. Only a small percentage of unlucky patients experience continuing problems. But it is precisely this minority that tends to dominate the forums.

Attention is what they seek, and technology is there to facilitate.

Take the website MDJunction.com—"Online Support Groups for Your Health Challenges"—which describes itself as "an active center for Online Support Groups, a place where thousands meet every day to discuss their feelings, questions, and hopes with like minded friends."

Participation in a community has been shown to encourage health and longevity and altruism is a known upside of online social interaction. So it's fitting that support and encouragement are the mandates and stated purposes of MDJunction, which claims on its homepage to have had more than 16 million visitors in its first eight years of operation. The site offers eight hundred different support groups where the ailing find their homologous counterparts—sufferers of lupus, rheumatoid arthritis, fibromyalgia, cirrhosis, Crohn's disease, various mood disorders, and many other maladies.

Not long ago, it would have been inconceivable—not to mention logistically impossible—to gather so many sufferers of elusive and complex illnesses in one place, particularly given the statistical rarity of some disorders, or connect them in real time. MDJunction does just that, offering a place for individuals to meet, gather tips on nutrition and diet, as well as share treatment advice. In an "off-topic" lounge, participants can tell jokes and let off steam.

It's a medical matchmaking service of sorts—syndicating individuals with similar symptoms. And the way people sort themselves into illness forums creates a fundamental shift in the psychology of the environment. For epidemiologists who study the incidence of illness in the general population, disease is a numbers game. For example, how many people do you know personally who've had a cold? Let's say hundreds or even thousands. How many people do you know who've had swine flu? Not many. This difference in incidence doesn't matter just to scientists. It informs our own perception of our health and vulnerability. In other words, due to relative incidence, you feel more likely to get a cold than swine flu.

Until you go online . . . where a virtual community, drawn from a population of hundreds of millions, is suddenly on your desktop, and nearly all of them with a similar complaint. This has a rather grave cyber effect on your perception. For Lisa, she was plunged into a world where every single person she would encounter or read about claimed to have Lyme—and often in some drastic late phase. This too, as discussed previously, is online syndication—the one-to-many and many-to-one interactivity facilitated by algorithms. It's similar to how deviants and swingers find each other online, but with a different result. In the environment of a medical forum, surrounded by fellow sufferers, you are much more likely to believe you are ill, and even seriously so.

The participants in the Lyme forum, where Lisa wound up, had raised more than twenty thousand different topics regarding the disease, and these had generated seven times as many replies. And despite the promises of comfort, most of these were terrifying. In 2009, a young mother shared the details of her treatment regimen with a Lyme newcomer:

> I am on 40 pills a day, b-12 shots twice a week and Rocephin IVs 2 times a day, four days a week. My LLMD is going to umm change some things up when I go back in June if I am not better by then.
>
> Only time will tell. Ummm I was diagnosed in January. I have been sick most if not all of my life.
>
> All we can do is learn as much as we can to fight this disease and it's co-infections and fight to get well. Just stay positive and in time, it will be easier to accept the diagnosis and start to move forward.
>
> At times it is overwhelming because I passed it to both my daughters and I don't like seeing them hurting. But I have ummm talked to them about the disease and they are ok with it. That makes it easier on me.

Who are these "experts" in the medical forums? In that lean, virtual, and socially impoverished medium, the old-school metrics can't be used to establish trust. You don't know whom you're communicating with, so—much more important—why are you taking their advice?

According to the forum, there was evidence from a new study that Lyme might be passed from a pregnant mother to an unborn child, and another study suggested that the disease could be sexually transmitted.

The fact that most doctors would consider these studies inconclusive was not mentioned. And like most of these sites, content is not monitored for accuracy. Medical misinformation can also leak and spread from one site to another, making it seem consistent and valid as you search, but in fact being merely the result of copy-and-pasting of erroneous facts and opinion. A Pew Internet Project study of medical information searches in the United States showed that eight of ten online health inquiries began with a search engine, but only a minority conducting them bothered to check the sources of the information or the date that the webpage was created.

Lisa's panic rose as she read the posts and comments from supposed Lyme "insiders," her mind swirling with all the new information about drugs, injections, symptoms . . . *Come to think of it, my neck does feel a little stiff*. If she had approached this physical sensation logically, she would have realized that her neck pain was likely the result of all the hours she'd just spent hunched over her laptop, not Lyme.

But health anxiety is not about logic.

Down at the bottom of the MDJunction webpage, there was a fine-print disclaimer, one of those endless scrolls of terms and conditions that nobody reads before they click "Agree." It reminds me of the notices in restaurant cloakrooms and restrooms: "We are not responsible for lost or stolen items."

These signs, it could be argued, indicate that the proprietor is aware of the risk by specifically denying responsibility for it. The medical disclaimer at the bottom of the page where Lisa was looking—"The information provided in MDJunction is not a replacement for medical diagnosis, treatment, or professional medical advice"—leads to the same idea.

But by then Lisa—formerly healthy, thoughtful, and relatively carefree—had descended into hysteria. As the sun began to rise, overtired and strung out, she became determined to not let Lyme disease overtake her body and ruin her life, as it had ruined the lives of so many she now felt she knew. She searched for a local medical clinic, checked the hours of operation, and emailed for an appointment, even though it was a "walk-in" clinic that didn't require one. She bookmarked the driving directions on her mobile phone, wrote down the dose of antibiotics she planned to ask for, and called her husband to recount the bad news.

Her visit to Cape Cod was a disaster. By day's end, when she finally reached home—having made no progress with Michelle in planning their high school reunion—Lisa had spent hundreds of dollars on an unnecessary doctor's visit, obtained a prescription for a course of strong antibiotics that would give her indigestion, gastritis, and an awful yeast infection. Worst of all, the chair in the doctor's waiting room where Lisa was sitting had been previously occupied by a teenage boy with a highly contagious viral flu and a 104-degree temperature. Three days later, Lisa came down with the same flu—and gave it to her entire family. They spent the next two weeks in bed.

Did Lisa have Lyme disease? Almost certainly not. Her malady was just as insidious. Overnight, she'd become a poster child for a twenty-first-century phenomenon called *cyberchondria*.

Symptom + Algorithm = Heightened Anxiety

The human mind is the most complex of machines, a wonderful labyrinth of biological engineering and design. But despite centuries of scientific investigation, the darker reaches of the psyche are still poorly understood. *Psychosomatic* involves both mind and body. There's a common expression that, while reductive, describes it pretty well: "It's all in your head."

Nobody has quite described the classic health anxiety sufferer as well as Joseph Heller. In *Catch-22,* his satirical novel about World War II, we meet Hungry Joe, an air force pilot who has completed his mandatory requirement of flying fifty missions and is now waiting to go home. Hungry Joe has quite an arsenal of neurotic inclinations and compulsions. He has "collected lists of fatal diseases and arranged them in alphabetical order so that he could put his finger without delay on any one he wanted to worry about."

The novel predates the Internet by decades, but Hungry Joe's condition is very twenty-first-century: fingertip search, categorized lists, a need to worry that only escalates with more searching. He's a high-achieving anxiety case addicted to worrying.

The term *cyberchondria* was first coined in a 2001 BBC News report, popularized in a 2003 article in *Neurology, Neurosurgery and Psychiatry,* and later supported in a groundbreaking study by Ryen White and Eric Horvitz, two pioneering research scientists at Microsoft who wanted to describe an emerging phenomenon engendered by new technology—a cyber effect, in other words. In the field of cyberpsychology, we define *cyberchondria* as "anxiety induced by escalation during health-related search online."

The word *anxiety* is really a catchall for a range of emotions that are a manifestation of anxiety—worry, nervousness, dread, panic, hysteria. The word *escalation* in this medical context refers to the common trajectory from run-of-the-mill complaints to more serious and rare illnesses. You search online for "sore throat," for instance, and find yourself engrossed and horrified by descriptions of esophageal cancer. Anxiety escalates as well.

How serious is cyberchondria? It's helpful to think of psychosomatic-type conditions as a continuum. The family of illnesses now called *somatoform disorders* all involve presentation of bodily symptoms suggestive of medical problems for which no organic basis can be found. At the far left, you have the "worried well," and that's most of us, who may imagine we are sick sometimes when we aren't. Next on the continuum, you have "health anxiety," which can be fleeting but is no fun. It is described by Gordon J. G. Asmundson, a leading scientist in the field, as being due to "fears and beliefs, based on interpretations, or, perhaps more often, *misinterpretations,* of bodily signs." Anxiety occurs in normal populations and in some cases can interfere with and disrupt relationships, employment, and recreation activities.

"Cyberchondria" comes next in seriousness, and it can also be fleeting or chronic. I first heard about it through the work of White and Horvitz, who published their research in 2008, starting the process with a 40-million-page random sample of "Web crawls."

Forty million? Now that's what I call a study.

Scientific research has relied heavily on sampling since 1662, when the English merchant John Graunt invented a method to estimate the population of London using partial information, that is, based on a sample. The very idea of a sample now seems pretty quaint. With the rise of big data, traditional research methods are being jettisoned. Why look at a limited subset or representative sample of a data set when you can look at everything? The unknown sample—*N*—that scientists seek to define could eventually encompass the entire known population itself.

White and Horvitz, in the largest study of its kind—can we stop for a second and try to fathom what it means to consider 40 million pieces of anything?—were able to support the relationship between medical search online and heightened incidence of health-related anxiety. When ten thousand of these Web crawls were manually analyzed, the research scientists were able to confirm there was an escalation in many searches—showing that searchers had progressed from reading about normal complaints to looking at rare and serious medical conditions. In a companion survey of 515 individuals, nine of ten respondents said they'd conducted at least one online search for symptoms of common medical conditions that led to a review of more serious illnesses. One in five said this happened to them "frequently."

The conclusion: "The Web has the potential to increase anxieties of people who have little or no medical training," White and Horvitz wrote, "especially when Web search is employed as a diagnostic procedure."

I met Eric Horvitz at a Microsoft summit in 2012 and sat beside him at dinner. At the time, probably less than a handful of people worldwide were researching cyberchondria. By then, I had spent three years studying his brilliant work, conducting research, and writing papers inspired by his initial findings. We talked nonstop that night about everything from hypochondria to Munchausen by Internet (which I will describe a bit later in this chapter). Our table companions looked on, baffled, as if trying to figure out what language we were speaking. "Cyberchondria?" one of them chimed in. "I haven't seen that movie yet."

The desire for information that will help us thrive is a natural, primal urge, as discussed in earlier chapters, like the act of seeking itself, a gratifying and dopamine-rewarding experience. But in the case of medical information, the good feeling doesn't last long.

Why? Because for years and years, the algorithms that drove search results did not take clinical incidence and statistical probability into account. Instead, they were based on advertising-type search models. Rankings were done by "frequency"—a crude metric that wound up presenting information very differently from the way a doctor would. Imagine for a moment that you went to your doctor complaining of a headache, and she said, "Well, you could have anything from a hangover to a brain tumor." "Oh my goodness," you'd likely respond, "talk to me about the brain tumor!"

Obviously your doctor wouldn't say that, but this is what health-related search was like. It offered a selection of options that included the worst-case scenarios, and people had a tendency to click on those first. This impacted the frequency rankings, moving the worst-case results up to the top of the results. And because the algorithm did not take into account your age, gender, general health, or medical history, it had little or no context.

Two of the leading causes of anxiety are uncertainty and perceived danger. The cyber effect of online medical search only adds information overload to the mix, creating a vicious cycle that may lead anybody into a vortex of dread. The bottom line: Instead of getting the considered, educated opinion of a medical doctor who has studied probability and risk, the worried individual looks at the search results and sees at the top of the list some of the most extreme, horrifying scenarios.

This is why a search would offer a brain tumor as a possible cause of your headache, without noting that this type of tumor occurred in just .002 percent of the population, mostly in a certain age group that was probably not yours. Yes, your headache could spell certain death. But the chances of that are infinitesimally small. The only thing that health search gave you was a big dose of anxiety, which is far more likely to harm you than the illness you don't have. It was like this from the inception of search until early 2015—when everything changed.

Now if you put in the words "head" and "neck" into a search, a pop-up box appears with a probability statement for "head and neck cancer." Yes, you may still be studying the symptomatology of a wide assortment of malignant brain tumors, but now, at least, you'll know the chances of having any of them are "rare."

But that won't deter some people from searching and searching, almost as if they are hoping to find something to worry about.

Hungry Joe + Hypochondria

Hungry Joe became anxious if he didn't have missions to fly. He had become addicted to the adrenaline rush, and he missed the anxiety and drama when it was gone. In the absence of that excitement, he looked for other things to worry about, and his alphabetical list of maladies and disorders provided it.

Moving along my continuum of health anxiety, we wind up on familiar ground, *hypochondria*—clearly the condition that Heller meant for Hungry Joe. Once upon a time in recent history, there was a genuine disorder called *hypochondriasis*. It referred to a situation in which individuals were inordinately preoccupied with their own health—or rather their poor health. I say "was" because the *Diagnostic and Statistical Manual of Mental Disorders, Fifth Edition* (DSM-5) made a number of sweeping changes. About 75 percent of what we previously called *hypochondria* is now subsumed into a new diagnostic concept called *somatic symptom disorder*, and the remaining 25 percent is considered *illness anxiety disorder*.

Losing an official diagnosis and description may cause anxiety in its own right. Many people become quite attached to their diagnostic labels and don't appreciate watching their illnesses classified away. In the interest of clarity, please forgive me if I just use the good old-fashioned term *hypochondria*, as it is popularly understood. All the literature that I will refer to, written over the past hundred years or so, uses that term.

What are hypochondriacs really like? They have a general preoccupation with health that is distracting and life-altering—and they are sure there is something seriously wrong with them. Rather than consciously faking their symptoms (like some other disorders, which I will get to soon), hypochondriacs can have authentic physical symptoms and changes in the body that cannot be traced to a cause other than the mind. Family members may roll their eyes—"Guess what kind of cancer Mom thinks she has now?"—but in many cases, Mom truly feels the pain, numbness, tingling, or tightness that she persistently complains about.

Fed by a great deal of anxiety and imagination, hypochondriacs have a deeply held belief that small symptoms and physical glitches are signs of a serious and probably fatal condition, but have trouble supplying a precise description of their symptoms. They are amateur medical experts, avid readers of health newsletters, medical studies, medical magazines, and books about medical topics. They repeatedly seek medical advice and make frequent visits to the doctor's office, but their anxiety and concern aren't abated by a doctor's reassurances that they are fine. In fact, they are likely to be disappointed when no actual physical problem is found. This only escalates their ruminations about an undiagnosed disease.

Another trait of the hypochondriac is a distrust of the medical profession—and a subconscious wish to undermine the authority of the doctor. The patient knows best. When a doctor can't seem to find a problem—and therefore no cure—the hypochondriac will move on to a new doctor. The hiring and firing of many physicians in an unreasonably short period of time is another indication of the condition.

You'd be hard-pressed to find many doctors who enjoy their time with these difficult patients. Some doctors regard people with health anxiety as nuisances who take up space and time that could be devoted to truly sick people in need of care. And when the day finally comes when a doctor gives a patient the news that they do have a diagnosable condition, and it's called somatic symptom disorder or hypochondriasis, this only leads to more frustration and upset, often causing a sufferer to become even more focused on health. This would be the time to seek the help of a psychotherapist, but I'm afraid that rarely occurs.

Hypochondria is found in 4 to 9 percent of the population, a figure that had been stable across studies for decades prior to the introduction and widespread use of the Internet. The condition appears equally among women and men, and seems to run in families, either due to a genetic predisposition for the disorder or behavior learned from their home environment. Like most individuals with these types of disorders, and not surprisingly, the hypochondriac experiences a wide range of emotional difficulties in life—often having trouble with impulsiveness, neurotic tendencies, and hyper-self-consciousness. There is usually an accompanying anxiety disorder or depression.

Prior to the introduction of the Internet, your good, old-fashioned hypochondriac had to painstakingly thumb through great tomes such as *Gray's Anatomy*—1,217 pages of obscure convoluted medical text—to provoke new ideas and fuel their anxiety. Now the obvious place for these frustrated folks is the Internet. You can find them hanging out in medical chat rooms, discussing their rare illnesses in forums, and logging on to intuitive diagnostic websites—where algorithms lead them, click by click, through the experience of being diagnosed with the same simplicity that prompts the general computer user through the installation of a new piece of software. (More on these "diagnostic" sites later.) Now all that information, and a great deal more, is available with the click of the keyboard, and about as hard to resist for the hypochondriac as Internet porn is for the pornography addict.

So what is the motive for anyone to engage in online medical search? My own research in this area found that many people are simply curious, like acquiring new knowledge, and enjoy the feeling of empowerment—which gives them confidence to challenge medical authorities and professionals. Others could not afford the cost and time involved in the medical consultation process. Another interesting piece of the motive puzzle: One researcher found that the average length of a consultation with a family practitioner is about eight minutes, but the "super-knowledgeable" patient who comes prepared with his search printouts tends to take up more of a physician's time. This is an interesting gain for the attention-seeking patient in the era of overworked physicians with a cattle run of patients: more face time with the doctor and value for your money.

Psychosomatic-type behavior is highly complex and often associated with impulsivity, which itself is associated with addictive behaviors. Intermittent reinforcement, as discussed in chapter 2, could also be a motivation for the cyberchondriac, as the Web symptom crawl becomes a form of lottery ticket that occasionally delivers a rewarding result.

Munchausen by Internet

Farther along the continuum, we finally arrive at *conversion disorders,* a category that represents some of the more extreme manifestations of chronic illness. In common parlance, these are "hysterical" conditions, which used to go by such names as "hysterical blindness" and "hysterical paralysis." Now renamed *functional neurological symptom disorder,* Munchausen syndrome, as it was originally known, is a psychiatric condition in which patients deliberately produce or falsify symptoms or signs of illness for the principal purpose of assuming the sick role.

Munchausen itself has a colorful background. Back in the eighteenth century, Baron Karl Friedrich Hieronymus Freiherr von Munchausen was a mercenary during the Russian wars of Peter II against the Turks. In retirement he became something of a curiosity—and renowned for telling tales of his adventures, often "over a bottle." In 1785, when Munchausen was still living, a scholar-turned-thief named Rudolf Erich Raspe published a book called *Baron Munchausen Narrative of his Marvellous Travels and Campaigns in Russia,* which made the most of the Baron's outlandish stories. A huge success, the book forever associated the name Munchausen with wild falsehoods. This led a British doctor, Richard Asher, to refer to patients who fabricated dramatic illnesses as suffering from "Munchausen syndrome" in 1951. And later, a pediatrician, Roy Meadow, borrowed the term to describe those who fabricated or induced sickness in others (usually their children) as suffering from "Munchausen syndrome by proxy" in 1977. Since then, there have been controversies about diagnosis and expert testimony, and a proliferation of various conditions that were once just called Munchausen by proxy.

People with a Munchausen-type syndrome, also known as "hospital addicts" in the U.K., are a much rarer bird than hypochondriacs, but they share a few traits, one being an antagonistic relationship with members of the medical profession. Like hypochondriacs, they also tend to be highly knowledgeable about symptoms, medical treatments, and care. As Laura Criddle, a critical care nursing specialist, relates in her article "Monsters in the Closet," quite often the Munchausen syndrome individual develops the condition after hospitalization for a true illness and then begins to lie pathologically about made-up maladies. They seem to be gratified emotionally by fooling doctors, and enjoy having a deceptive relationship with them.

When the syndrome takes a different name, Munchausen by proxy, it is rarer still—and the motivation changes. The individual with the syndrome (most often a mother, in 93 percent of the cases) uses another person (usually the mother's child) in order to work out a disturbed scenario with medical experts. Dr. Meadow, who gave the behavior its name, described one mother who mixed her own blood into her baby's urine sample in order to fake illness. Another mother poisoned her toddler with table salt.

One of the most famous cases of MSbP, as it's called in the clinical journals, is that of Kathy Bush, a Florida woman whose twelve-year-old daughter, Jennifer, endured two hundred hospitalizations, forty surgical procedures, multiple poisonings, and treatment for dozens of serious and suspicious infections over the course of her childhood. After Kathy Bush was arrested in 1999, the girl was sent to live in foster care out of state, where she received medical care and improved. Later, she was reunited with her mother and denied abuse had taken place.

It's a form of child abuse, but unlike a parent or caregiver who cruelly harms children by lashing out in anger and frustration, the Munchausen by proxy mothers are motivated by an insatiable need for social attention and recognition. They become emotionally rewarded by their ability to fool people, to receive praise and sympathy, and to be perceived as being embroiled in something important. Of course, there are sometimes secondary gains or motives—subsidized housing, welfare, free medications, and financial aid. But the big driving factor seems to be a desire to be seen as an angel, a rescuer—a stoic caretaker of a chronically sick child. It becomes the mother's identity.

Technology has the power to exacerbate a factitious disorder—when a person deliberately feigns or exaggerates having an illness. It doesn't require a huge imagination to see what happens when you mix a disturbed individual who likes to fake illness with the richness and plausibility of information now available online—or the communities that gather there. As long ago as 2000, Dr. Marc Feldman, a psychiatrist in

Birmingham, Alabama, who specializes in factitious disorders, coined the terms *virtual factitious disorder* and *Munchausen by Internet* to describe those who use the Web to aid their deception and masquerade. (Feldman himself runs a Munchausen website, where members of the cyber-public can report new suspicious cases.) Some specialists have even made the case that Munchausen by Internet is a natural evolution of the original condition—a more efficient and rewarding environment for the patient's deceptions. Think about it. Logically the sympathy of hundreds of people online is far more powerful than sympathy from one person in a white coat. And rather than deceiving a small circle of friends, neighbors, and medical professionals, they can fool potentially millions.

A shocking example of Munchausen by Internet was perpetrated by a blogger, David Rose (a.k.a. Dave on Wheels), who described himself to his virtual audience as a profoundly deaf cerebral palsy victim and quadriplegic living in Los Angeles. Over a period of four years, between 2008 and 2012, he inspired a growing fan base of followers on Twitter and Facebook with his funny asides and heartbreaking story, which he said he wrote using a Tobii computer that followed the movements of his eyeballs. As he garnered more and more attention, Kim Kardashian and other celebrities began retweeting Dave's sad musings and bravely uplifting one-liners. Then, out of the blue, his "sister" Nichole posted the stunning news of Dave's death from pneumonia on his Facebook wall. This unbearable end to the Dave on Wheels saga drove thousands to leave "RIP" comments on his page while continuing to share his poignant story with others online. One die-hard fan even booked a flight to L.A. to attend Dave's funeral, prompting a confession from "Nichole Rose" that Dave was a fictional character, a persona invented to "inspire people to love and live a better life."

What drives an individual to such lengths of pathological lying and stagecraft is likely true mental illness—not just the desire to create an Internet hoax or cute joke. And in spite of Munchausen sufferers' need for sympathy, it's pretty hard to muster much of it for people who enjoy manipulating and fooling others, or the Munchausen by proxy individuals who abuse their children for emotional gain. This isn't usually the case in *cyberchondria by proxy,* the condition we'll get to next, which is quite often the result of good intentions gone very wrong.

Cyberchondria by Proxy

The *Sydney Morning Herald* told the story of a devoted son who was worried about his father, a seventy-year-old man who complained of body aches and jaw pain when he chewed. He had been diagnosed with temporal arteritis, an inflammation of the vessels that supply blood to the brain and head, and was taking a number of medications.

The son was suspicious about all those drugs—and wondered if his father's pain could be a side effect of one of his many prescriptions. Sure enough, he found exactly what he was looking for online. His father's symptoms were listed as possible side effects of Lipitor, the popular cholesterol-lowering medication that his doctor had prescribed. Well, that was easy. All his father had to do was stop taking it for a while and they'd see if his pain went away.

But when the man's father went off Lipitor, instead of getting better, his condition deteriorated rapidly. The pain grew more intense, and he was rushed to the hospital. After an examination by his physician, the man's temporal arteritis was found to have gotten much worse. If untreated, he would lose his sight. The reason he'd gotten worse? The Lipitor wasn't hurting him, as the online information seemed to suggest—it was keeping him alive.

In the newspaper account, the man's physician, Dr. Brian Morton, former president of the Australian Medical Association in New South Wales, spelled out the problem: "There is potential for Dr. Google and well-meaning family members," he said, "to cause catastrophe."

When reading the initial study of cyberchondria done by White and Horvitz, I was impressed with the attention to detail, the volumes of data, and the meticulous way that the two scientists had compiled their findings and supported the construct of the condition. But something else drew my attention: While the results showed that 58 percent of individuals who did a medical search online were searching for themselves, the figure that jumped out at me was the staggering number of slightly more than 40 percent who searched for relatives, friends, work colleagues, and practically anybody else.

Academics call this a "gap in the literature." I saw it as an opportunity to identify a new cyberchondria-type phenomenon, describing those who habitually and compulsively search for others. I used Munchausen and Munchausen by proxy as models, and after the exciting results came in and my findings were written up, the term *cyberchondria by proxy* was introduced.

Nobody would argue that responsibilities of caring for young children and elderly parents aren't sometimes overwhelming. When a child becomes sick with a high fever—even with the common cold or flu—parents experience an extreme kind of anxiety. According to Laura Criddle, all parents "present somewhere on a continuum of medical neediness." Many normal parents worry—even becoming hypervigilant or misinterpreting and exaggerating their own children's behaviors and symptoms. If by some chance you happen to experience a "near miss"—when a doctor or medical expert misdiagnosed your child's illness—this can also result in a growing mistrust of members of the medical profession. This kind of trauma can cause an individual to become more dependent on online searches, and one of the most unfortunate outcomes of cyberchondria by proxy is the amateur prescribing of cures and medicines, the dispensing of prescription medications to unsuspecting friends and family. Don't share your pharmaceuticals with loved ones! And don't send anyone to a so-called Canadian pharmacy to buy cheaper drugs. In 2013, more than 1,600 of these Internet pharmacies (none of which was actually in Canada) were shut down. In many instances these drugs, when purchased by undercover agents and tested, were found to be not cheaper generic drugs but out-and-out counterfeits.

My own research findings in this area included some cautionary tales. One woman reported an incident in which her partner had been given medication by a relative who had searched online, and the dose provided had been too strong—in fact, it was an overdose. A number of participants described bringing printouts to a doctor's office, referring to a Web search as similar to getting a "second or third opinion," and noted that the practitioner's attitude toward their self-diagnosis was "disdainful" or "dismissive" or "irritated." Participants were supportive of using chat rooms for medical information exchange.

Interestingly, cyberchondria by proxy individuals do exhibit behaviors found in some of the more serious disorders—the challenge to authority, compulsive medical information seeking, escalation, and symptom checking. Connecting these dots was beyond the bounds of my study, but it may be an important component of fully understanding this phenomenon in years to come.

One finding spoke volumes to me: Participants with a medical family background reported resisting the temptation to search.

Your Imagination + Artificial Intelligence

There is regular search—just plugging in a word, or two words, and getting results that make you anxious. That's where most people start, hoping to satisfy their curiosity about a medical problem.

Next, they might visit a diagnostic website. These sites are designed to lead a worried participant through a set of questions to determine a probable diagnosis.

This is where the power of suggestion becomes a factor. As humans, we can be highly suggestible. Suggestibility describes someone who is "easily influenced by other people's opinions," or, in this case, the opinion of an algorithm.

How can a normal, healthy person who's experiencing some health anxiety wind up manifesting some actual symptoms?

Let's say you have a pain in your arm that you're worried about. You did a pretty exhausting workout in the gym yesterday, but that fact isn't front and center in your mind. So you find a nice diagnostic website to ease your worries. If you enter the words to describe your main concern, "pain" and "arm," you will be presented with a nicely laid out graphic with little directional arrows. First you'll be asked if you've experienced pain in your arm. When you answer yes, you'll next be asked if you're experiencing tingling in your fingers.

Here's something to try at home: Hang your arm down by your side, relax, close your eyes, and wait a minute. If you imagine a line going from your tired neck to your stiff shoulder and through your sore tennis elbow all the way to the tips of your fingers still pulsing from relentless tapping on keyboards eight hours a day . . .

Do you feel that tingling?

Of course you do.

Now you've graduated to the next question on the diagnostic website: "Is the pain in your arm radiating across your chest?"

Now that you mention it . . .

You click the box that says "Yes."

Question #4: "Are you palpitating?" Of course your heart is beating fast—you are having an anxiety attack—but along the way, you have picked up a perfect "cluster of symptoms," something you've been taught by the website.

A real doctor has been taught to carefully avoid prompting patients with questions like "Is there tightness in your chest?" And medical professionals know that practicing good medicine means not encouraging suggestibility. That's where diagnostic websites fall short. They seem almost designed to spark the imagination. And while they attempt to provide diagnoses, they are actively suggesting, prompting, and even provoking symptoms.

Think about it this way: If you are having a problem with starting your car, the act of searching for an answer online is not going to intrinsically damage your car. But if you are having a problem with your health, the very act of reviewing symptoms may help them to manifest physically or mentally in a technologically amplified *psychosomatic effect*. I have another unofficial term for this, *technosomatic effect*.

So your heart is racing, your chest is tight. With your painful arm, you pick up the phone and call 911. When the EMT arrives, you reel off a perfect cluster of symptoms that have been prompted by the website. You may be the wrong age, have no relevant history, and appear reasonably well. But how can the EMT—or the emergency room doctor—be certain? Your perfect cluster makes it sound like a textbook cardiac event.

A medical professional has no recourse but to run a battery of tests, scans, EKGs, MRIs, and blood panels—or whatever intrusive diagnostic procedures are called for, given your cluster. If nothing unusual is found, often another round of tests will be ordered, and panels, scans, and follow-ups.

The irony of this sequence is that many of the diagnostic procedures carry an inherent risk. As anyone who has been brought back for a second ultrasound or a more comprehensive mammogram knows, one of those risks is the heightened stress and anxiety from the medical testing alone. Empirical evidence supports this. One of the highest anxiety-inducing events in life is awaiting the results of medical tests and diagnoses.

But something much worse can happen: A medical problem of *some* kind will be found. Why is that a problem?

Knowledge, as you'll see below, is not always a good thing.

Getting That Full-Body Scan

I remember when, in the early 2000s, my friends abandoned the purchase of the latest fashion accessory or beauty treatment in favor of the ultimate makeover: the deluxe "full-body scan." This was the start of a new wave of medical technology that allowed further exploration of the previously unknown. We were suddenly able to see—in great detail—what lies within us. And even better, we could be watching. Live.

From my perspective, many a good dinner conversation has been ruined by an enthusiastic blow-by-blow, or polyp-by-polyp, description by someone on the receiving end (literally) of a live stream colonoscopy. Never before have we been able to see inside ourselves so clearly! The procedure itself is all well and good—and has surely saved thousands of lives. But professional associations have warned that the scans lead to rounds of tests of innocuous lumps while missing common cancers.

Dr. Harvey Eisenberg was a cardiovascular radiologist who took an aggressive approach to wellness by performing comprehensive full-body scans on the "worried well" at his Health View Center for Preventive Medicine in Newport Beach, California. His motto was "You don't know what's inside until you look," and his "preventive" scans created quite a media sensation. Forget the colon; Dr. Eisenberg wanted a look at every organ, every bone, and all tissue. His approach struck many as cutting-edge medicine. He was celebrated on *The Oprah Winfrey Show,* and consulted, and endorsed by the likes of William Shatner and Whoopi Goldberg. Between 1997 and 2000, more than fifteen thousand individuals were scanned from neck to pelvis at the Health View Center in hopes of discovering and preventing any nascent illness. Studying the results, radiologists looked for early signs of heart disease, cancer, arterial plaque, lung nodules, inflamed prostate, and spinal vertebrae degeneration.

Guess what happened?

An evolving pathology was found in *every case.*

Every single one.

Think about that.

Dr. Eisenberg claimed that he'd never seen a normal scan. "It's a daily event for us to uncover unsuspected life threatening disease," he said, "that is either stoppable, curable, or reversible."

The radiation exposure during whole-body CT scanning is low, but still a concern. When asked about this, Dr. Eisenberg said, "Yes, there's an increased risk, but you don't get something for nothing."

A potentially bigger problem than radiation awaited the patients of Dr. Eisenberg. The bottom line: If every physical problem or evolving pathology inside our bodies were discovered and treated, would it save more people than it killed?

Iatrogenesis is a Greek term derived from *iatro,* or "healer," and *genesis,* meaning "brought forth," which refers to an illness "brought forth by the healer." It can take many forms: an unfortunate drug side effect or interaction, a surgical instrument malfunction, physician negligence, medical error, pathogens in the treatment room, or simply bad luck. An early study in 2000 reported that it was the third most common cause of death in the United States, after heart disease and cancer.

You heard that right. *The third most common cause of death.* Having an unnecessary surgery or medical treatment of any kind means taking a big gamble with your life. Before long, the U.S. Food and Drug Administration began issuing warnings about the potential risks of the scans, and health insurance providers not only refused to cover them but advised against them. When last heard of, Dr. Eisenberg and his full-body scanner were operating out of a van in Irvine, California.

Nobody wants to die before his or her time. And it turns out that worrying that you might be dying, and acting on that worry, can become a self-fulfilling prophecy. The full-body-scan boom went bust. And the worried well moved on—mostly online.

A Case for Death by Cyberchondria

We have to thank John James, a toxicologist at NASA who lost his son to what he believes was a careless medical mistake in a Texas hospital in 2002, for shedding light on the growing number of deaths from the iatrogenic effect. An estimate of between 44,000 and 98,000 deaths annually in the United States as a result of medical error in hospitals was reported in 1999, when the Institute of Medicine issued its now famous report, *To Err Is Human*. A decade later, the Department of Health and Human Services reported that about 15,000 hospitalized Medicare beneficiaries per month experienced an event that contributed to their deaths—that is, failures in hospital care. This results in an estimated 180,000 patient deaths annually. In 2013, with the help of James and the advocacy group he started, Patient Safety America, a stunning jump in deaths was reported: between 210,000 and 440,000 hospital patients in the United States die each year as a result of a preventable mistake.

By his reports, these deaths increased fourfold between 1999 and 2013. If anything, given the improvements in medical technologies in those fifteen years, the death rate should have decreased. Is patient care getting so much worse? Were the previous studies done poorly?

What is the difference?

James has been a tireless campaigner on the subject of healthcare for nearly fifteen years. His latest book, *The Truth About Big Medicine,* discusses the flaws of the American healthcare industry, reveals unsafe medical practices, proposes ways to correct injustices, and includes discussions of imaging, medical devices, pharmaceuticals, hospital procedures, and medical negligence.

While these dangers inherent in the American healthcare system may be responsible for a rise in the iatrogenic death rate, I would argue that another sort of effect may have intervened. The 1999 figure reflects a time prior to the prevalence of Internet medical search, and the escalation reported in 2013 more closely reflects the times we are living in now, when as many as 60 percent of Americans search for health information—35 percent of those search for diagnostic information, and just over half proceed to make an appointment to see a physician. It cannot be coincidental. This may be a pipeline: Online health-related search leads to escalation in anxiety, which leads to unnecessary medical visits and unnecessary medical procedures, which in turn leads to an increase in iatrogenic deaths.

A new study conducted by patient-safety researchers at Johns Hopkins University School of Medicine and published in May 2016 reemphasizes the seriousness of this issue. The findings show that medical errors in hospitals and other health-care facilities are "incredibly common and may now be the third-leading cause of death in the United States—claiming 251,000 lives every year, more than respiratory disease, accidents, stroke and Alzheimer's."

If I take a few steps back, what I'm observing is a deadly cocktail: increased media coverage of health and medical stories that drives anxiety, TV commercials pushing pharmaceuticals that target health uncertainty, more medical testing to protect doctors from medical malpractice lawsuits, and then you can add in the cyber effect, online search.

This is something Yahoo, Bing, and Google surely know about and have been grappling with for some time. Just a few years ago, when you did a search and used a medical or physiological term, a disclaimer popped up, saying that this was "not a diagnosis." This box disappeared, and then returned in the form of a probability statement. More recently, search sites have experimented with connecting medical searchers with actual doctors. I would like to think this is a positive step, in terms of an informed approach, and not just another way to monetize the miseries of the worried well.

While we can't know with certainty how many lives could have been spared over the past ten to fifteen years, we should ask this question: What would it take to create safer and more intelligent search protocols for medical queries? What would it take to rethink media coverage and TV commercials so they didn't fuel health anxiety? And what about reforming medical malpractice protocols to discourage unnecessary testing?

The Internet offers so many advantages, yet with the emergence of cyberchondria and other technology-facilitated problems appearing on the horizon, we have to hope there's a way for medical ethics to be translated online.

The Google motto "Don't be evil" reminds me of a line often attributed to the Hippocratic oath, *primum non nocere*. This moral code for physicians translates as "First, do no harm." It is something advertisers, media, online search engines, health websites, and the legal community may want to embrace.

Conclusion

Imagination is a beautiful thing, but it can be fierce and unruly. We would be smart to be respectful of its power. The fearmongering and hysteria over the Ebola virus in 2014 is proof of how health anxiety, and health scares, can cause more problems than the illness itself. In a terrible cyber twist, the scammers who sent out fake emails from the World Health Organization, using subject lines like "Ebola Survival Guide" and "Ebola Outbreak Now WORSE Than We're Being Told," managed to dupe thousands into surrendering control of their computers. Of course technology can have positive effects on healthcare and disease outbreaks. Recently, rival researchers studying the mosquito-borne Zika virus found a way to collaborate via Twitter.

Knowing everything about your body, your medical health, and the drugs you are taking is now de rigueur. But there's a lot of evidence to support my view that you can know too much. This finds agreement in Japan, where a doctor will rarely, if ever, tell patients that they have cancer. And under no circumstances are they told their condition is hopeless. It is commonly believed that patients who are told they are terminal tend to die sooner than those who aren't.

But in the West, instead of less information, we have more and still more—now made possible by search technologies. It wouldn't surprise me if there was a cultural change of heart about this eventually. In recent years I've heard stories from women with breast cancer and other serious diagnoses who decided not to share this information with friends and extended family, lest their own young children find out and go online to search outcomes. Even if there is no transparency about how medical-search algorithms work, we know where it leads. The young and the vulnerable, now exposed to a range of morbid and serious content that they cannot process or understand, are the ones we need to protect most.

Here's my own motto: *caveat inquisitor.*

"Let the searcher beware."

CHAPTER 8

What Lies Beneath: The Deep Web

Pirates and buccaneers have been romantic fixtures in children's literature and Hollywood movies over the past century. They are often depicted as courageous figures—swashbuckling thieves who roam the high seas with expert skill, defying governmental authority and living by their own "code of honor." In the language of storytelling, this code signals a sense of morality and makes the pirates human and almost admirable—the bad-but-good guys we can love.

Formal or informal codes of conduct actually exist for all groups, whether they are a family of primates or an underworld gang. Rules make everything work better. That's why it's called "organized" crime.

Most of us do not commit felonies in real life, or want to, but it can be fun to imagine living the lawless life of a pirate, navigating uncharted tropical waters and meting out our own brand of high-seas justice in the clash of a sword fight. Lawlessness mimics freedom, and the vicarious experience of watching it in a movie can be exhilarating. The golden age of piracy (about 1680–1730), as portrayed in the Pirates of the Caribbean films—with all those beautiful shots of blue water and desert islands, not to mention Captain Jack Sparrow played by Johnny Depp—is irresistible entertainment. It's easy to see why the very word *pirate* does not register as pejorative.

Where am I going with this?

The Deep Web is surely like the Caribbean of the seventeenth and eighteenth centuries, a vast uncharted sea that cybercriminals navigate skillfully, taking advantage of the current lack of governance and authority—or adequate legal constructs to stop them. And like the golden age of piracy, we are living in a time of upheaval and geopolitical changes both in real-world and cyber contexts, which can encourage a free-for-all of opportunism and a rise in lawlessness.

On the high seas of the Internet, secret hiding places for cybercriminals abound—safe harbors, concealed caves, and digital coves where they can sail in, drop anchor, and buy and barter for tools, weapons, and other contraband. For those needing accomplices, the bawdy pubs and tavern inns of the Colonial period have been replaced with covert forums, chat rooms, and criminal-networking sites, all of which are plentiful on the labyrinthine Deep Web. While you can find cybercriminals anywhere on the Internet, they have a much easier time operating in the murky waters of the darkest and deepest parts.

And just as the legitimate digital markets like Zappos and eBay have become more sophisticated about merchandising and sales, the online black market has grown increasingly efficient. I was a contributor to the Europol cybercrime threat assessment report in 2014, which described the emerging but hard-to-believe phenomenon of *crime as a service online*. Almost any kind of criminal activity—extortion, scams, hits, and prostitution—can be ordered up online now, thanks to well-run websites with shopping carts, concierge hospitality, and surprisingly great customer service.

This window of opportunity for cybercrime won't last forever. It can't. But until more laws to govern the open waters of the Internet are implemented and jurisdictions begin to coordinate across virtual borders—the way they do for sea and aviation laws—we will continue to live in a golden age of digital piracy, resulting in a lawlessness that impacts all of us.

Tip of the Iceberg

If you aren't exactly sure what the Deep Web is, you aren't alone. This part of the Internet is commonly discussed in law-enforcement and cyber-security circles, and is a source of fascination for the tech industry and media, but like so many aspects of the cyber environment, it has a feeling of being a "tech experts only" subject, as if it's too complicated for anybody else to fathom. When I'm on the road, whether I am speaking at conferences or discussing my work with people outside digital forensics, I am always stunned by how few people really understand this hidden part of the Internet or how it works.

The field called *cyber-security* is vast and pretty complicated. Non-experts are puzzled—and left with only one sticky mantra in their heads: *Change your password.* Well, that turns out to be great advice. But how is anyone supposed to remember all of those codes, not to mention the new ones?

I like to say that your memory may be weak, but your passwords need to be strong. That means that you should avoid taking the easiest route of simply changing one character of your password, like adjusting the ending from a "7" to an "8." Human beings are predisposed to gravitate to variations on a theme, due to the way human memory works. The ability to recall often is by association. Therefore, what seems clever and easy for you won't be enough to stump an identity thief. Let's not forget, one of the most impossible codes of all time, the Enigma code, was broken due to a similar mistake. Its Nazi operators got lazy. Instead of resetting the dials randomly of the Enigma machine every day, some operators changed a dial by only one notch, creating an easy pattern to crack.

Failed morals and antisocial personality traits aren't the only things that give cybercriminals an advantage over their victims. These con artists are expert observers of human behavior, especially cyberbehavior—they know how to exploit the natural human tendency to trust others, and how to manipulate people so they give up confidential information, or what is called a *socially engineered attack.* When it comes to identity theft or cyber fraud, it is much easier to fool a person into giving you a password than it is to try to hack it (unless the password is really weak). This sort of social engineering is a crucial component of cybercriminal tactics, and usually involves persuading people to run their "free" virus-laden malware or dangerous software by peddling a lot of frightening scenarios (which is why it's called *scareware*). Fear sells.

The quagmire of cyber-security is made only more complicated by all the interested parties trying to solve our problems. There's an industry of professional and amateur specialists promoting anti-malware, anti-spyware, and antivirus software—while flooding our in-boxes with a steady stream of cyber-security tips, delivered on every platform from blogs to YouTube. Due to the constantly evolving technology and information overload, the subject is now more mind-numbing than scary. Many people prefer to buy malware protection and just hope for the best.

But is that enough? You don't have to look far to find somebody who has been a victim of a cybercrime—whether it's credit card fraud, cellphone hacking, data hacking, data destruction, or cyber-extortion. Identity theft is now the number one consumer complaint at the U.S. Federal Trade Commission, and the U.S. intelligence community ranks cyber threats as one of the top global security concerns, along with terrorism, espionage, and WMDs.

So where are these cybercriminals hanging out?

The Deep Web. At one of my first meetings with CBS executives and the creative team of *CSI* to discuss my work in forensic cyberpsychology with law enforcement agencies worldwide, this mysterious place was one of the first things I had to explain. The *Deep Web* is often a misused term.

So what is it?

Simply put, the Deep Web refers to the unindexed part of the Internet. It accounts for 96 to 99 percent of content on the Internet, vastly larger than the Surface Web, where regular traffic is occurring. Most of this content is pretty dry stuff, a combination of spam and storage—U.S. government databases, medical libraries, university records, classified cellphone and email histories. And, just like the Surface Web, it is a place where content can be shared.

What makes it different? The content on the Deep Web can be shared without identity or location disclosure, without your computer's IP address and other common traces. The sites are not indexed and therefore not searchable if you're using a Surface Web browser like Chrome or Safari or Firefox. For software that protects your identity, an add-on browser like Tor is one of the most common ways in. The name Tor is an acronym for "the onion router" because of the layers of identity-obscuring rerouting.

Standing before the *CSI* team, as I prepared to describe the Deep Web, I told them to picture (as I do) the entire Internet as a giant ball of colored twine. "Only certain strands can be followed, but some cannot," I said. "The strands of twine that cannot be followed are in the Deep Web."

Looking around, I saw a room of perplexed faces. The visual metaphor that worked for me wasn't working for anybody else.

Strands of twine?

What is she talking about?

Trying another approach, I walked to a whiteboard on the wall and drew a long horizontal line. "Here's the horizon of the ocean," I said. Then I drew the tip of an iceberg rising up from the ocean's surface.

"The tip is the Internet you know—the one that you regularly access and browse," I said. "It seems unimaginably big—doesn't it?—like an entire universe of information."

Heads nodded.

"But content-wise, it's actually pretty small. What lies beneath is the Deep Web. It's almost one hundred times bigger."

The room grew quiet. Trying to fathom the size of the Deep Web is kind of mind-blowing and takes a while to process. When I was a kid, I used to spend hours thinking about the universe—struggling to comprehend the concept of infinity. I used to think about it until my brain hurt, hoping that eventually, if I thought long and hard enough, I would understand how to quantify it. But sometimes the best path to understanding is to grapple not with size but with characteristics. The universe is truly limitless, but that doesn't keep us from being able to imagine a solar system or even a galaxy. What helps is creating a visual analogy, like twine, an iceberg—or the endless blue waters of the Caribbean in 1699.

So let's discuss the characteristics of only one aspect of the Deep Web, where a lot of cybercrime operations are run, where the pirates sail and loot and bury treasure. It is called, somewhat romantically, the Darknet or Darknets. It is only a fraction of the Deep Web, and a tiny fraction of total Internet content. In network science, the word *dark* is used to describe anything that is hidden, untraceable, and unfindable. When a spy goes dark, he can't be located; when you go dark online the same thing happens. Darknets refer to what is deliberately hidden. Quite a lot of what happens there is illegal.

How did illegal operations wind up there? The Deep Web was first used by the U.S. government, and the protocols for the browser Tor were developed with federal funds so that any individuals whose identity needed to be protected—from counterintelligence agents to journalists to political dissenters in other countries—could communicate anonymously with the government in a safe and secure way. But since 2002, when the software for Tor became available as a free download, a digital black market has grown there, a criminal netherworld populated by terrorist networks, criminal gangs, drug dealers, assassins for hire, and sexual predators looking for new images of children and new victims.

As explained in one Darknet tutorial:

> There are things like blogs, forums (from normal to revolutionary to blatantly illegal), Tor-enabled instant messaging and chat, anonymous file hosting, anonymous financing, anonymous tipping and information exchanges, information on computer security/anonymity, info on warez/cracks/hacking, all the books, music, movies you can possibly imagine, even links to sports betting and trade information, links to international drug markets, prostitution rings, assassin markets, black market products, [child abuse material]. Some of societies [*sic*] most deviant people use this network. Not just those that browse the sites on there but also those who create and manage them.

Some call the entry point or portal to Darknets "the gates of Hell," because once you pass through, there's no telling what you'll encounter.

No exaggeration: Almost anything you can imagine or have seen in a Quentin Tarantino movie that's illegal, illicit, and contraband can be purchased on Darknets. This has led to flights of fantasy and myth about what exactly goes on there, because the whole thing seems so impossibly lawless and surreal. One of these Darknet rumors is the existence of fighting tournaments, Roman gladiator–style, where brutally violent matches are live-streamed for high-paying customers. The matches are said to be skillfully produced and pit two professionally trained fighters against each other.

As *Thebot.net* explains:

> It may seem surreal but they are guys that train with the best and want no part of UFC or any fight league. Dudes who really enjoy fighting to the death . . . it's not some barroom brawl. These things happen and a lot of millionaires pay big money to see them. Modern Gladiator battles. I heard there are some with humans vs. animals.

To the death? *Can that really be true?* In my work, I've had to become familiar with some dark corners of the Deep Web. The scope of human depravity there is unfathomable and deeply disturbing. But to visit for fun, for entertainment, on a dare—or even out of curiosity—is a serious mistake with potentially lasting repercussions. This isn't HBO. It's real. And can be dangerous.

This brings me to the next question I'm always asked: How can illegal operations thrive in the Deep Web? Isn't there a way to effectively monitor and police the space?

Size and scope is a problem. There is an almost infinite number of hiding places, and most illegal sites are in a constant state of relocation to new domains with yet another provisional address. Another matter that confounds law enforcement is that many of these sites do not use traceable credit cards or PayPal accounts. Instead, virtual currencies such as Bitcoin are traded.

Bitcoin is the solid-gold doubloon of the digital realm—only better—

it's untraceable, anonymous currency, or what law enforcement calls a *cryptocurrency.* Secret money, in other words. The cyber equivalent of unmarked bills. While it is traded by reputable entities for totally legal purposes, so far the existence of cryptocurrencies in general has not been a particularly positive thing for law enforcement. The effects of anonymous money on human behavior are similar to the effects of anonymity online in general. It can amplify and facilitate certain behavior—in this case, covert behavior. The ramifications are enormous.

Historically, the biggest problem with crime was its potential to disrupt social order and encourage more disruption. Violence begets more violence, theft negates toil, and victims take time to recover—even in prehistoric societies. In the twenty-first century, in a Western and liberal democratic context, notwithstanding centuries of legal code and a technologically advanced law enforcement, society appears to be in the midst of a wholly new forensic experience.

Crime as a Service

Just as real-world crime has been impacted and facilitated by the cyber environment, the society of criminals online has been influenced by the Surface Web marketplace and its services—from Uber to Tinder to Amazon. There is recognition that consumers now expect instant gratification as well as bargains. The Darknet sites selling illicit drugs, weapons, and hidden services offer discount days, coupon codes, two-for-one specials, money-back guarantees, and loyalty points. Promotional campaigns are common, and some drug-trading sites offer escrow services; they will hold your money until your package arrives safely. When there's a technical glitch, they are quick to apologize.

There's even after-sales follow-up. In Europol reports, we describe it as "Crime as a Service Online" (CaaS). Customers are asked about their level of satisfaction and given opportunities to offer suggestions for improvement. Of course, it's one thing if your local car salesman knows your address and can swing by if you have a problem with your latest purchase. But it's an entirely different matter if organized cybercriminals try the same thing.

Anonymity works both ways. The same mechanism that provides

anonymity to users provides a cloak of invisibility to the hosts of criminal sites. This has given rise to what law enforcement calls "hidden services," the most prevalent of which is the selling of stolen credit information, or *dumps*. The competition is so stiff that these sites have their own service-minded rules to keep customers happy, frequently offering refunds for stolen cards that are declined at a retailer. *No risk to you!*

McDumpals, one of the leading sites marketing stolen data, has a clever company logo featuring familiar golden arches and the motto "i'm swipin' it." And the McDumpals mascot, a gangster-cool Ronald McDonald, points a handgun at the viewer. So cute.

Customer-friendly interfaces and glib ads can even be found on sites where assassination-for-hire is offered. One site boasts, "I always give my best to make it look like an accident or suicide." Another says, "The best place to put your problems is in a grave." These services sometimes offer horrific bonuses, such as a chance to win a little money back with a fun wager: Followers can gamble on the actual time of execution. No doubt, this takes the science of digital marketing to a whole new level, and it's a disturbing one.

Which brings me to the story of Ross William Ulbricht. You may have already read about this young man from Texas who had so much going for him. Nobody doing business in the Darknet knew him by his real-world name, though.

He had an alias: Dread Pirate Roberts.

The Silk Road Story

Ross Ulbricht was born outside Austin, Texas, in 1984. He was a smart, conscientious Boy Scout who loved comic books, skateboarding, and math. According to his father, Ross was "a healthy, happy, unflappable Buddha of a kid." Spending summers in Costa Rica, where his parents, both entrepreneurs, built and rented bamboo solar-powered houses, Ross ran barefoot, learned to surf, and proved to be "a little too fearless," according to his mother. He loved nature and being outdoors.

"We didn't want our kids on the computer," his father told *Rolling*

Stone in a long and fascinating profile. "We wanted them outside playing."

In high school, Ross devoted much of his free time to drug experimentation and parties, but managed to do well enough in school to receive a scholarship to the University of Texas at Dallas. There, he published papers on solar-cell technology at the NanoTech Institute, and was remembered for not wearing shoes or a shirt on campus and for his love of psychedelics and Eastern philosophy.

"My whole philosophy at the time," Ulbricht recalled later, was "of being super-open and loving and connected to everything." One college friend described him as a "physics hippie."

At Penn State, he studied for a master's degree in materials science and engineering on another full scholarship, and got into yoga, conga drumming, and the writings of Austrian economist Ludwig von Mises, who espoused the virtues of the free market. On Facebook at the time, he enthused in a post, "overwhelmed with the glory of being alive."

Becoming disillusioned with academia and science by the time he finished his master's in 2009, he changed his life plans. "He wanted to be an entrepreneur," his mom told *Rolling Stone*.

Bitcoin had just become available by free download. For Ulbricht, a true believer in free markets, cryptocurrency must have felt like a gift from the tech gods. In a journal that was later found by the FBI, he wrote: "Every action you take outside the scope of government control strengthens the market and weakens the state." He described wanting "to create a website where people could buy anything anonymously, with no trail whatsoever that could lead back to them."

Silk Road was an online black market, the first of its kind—offering drugs, drug paraphernalia, computer hacking and forgery services, as well as other illegal merchandise—all carefully organized for the shopper. And like Amazon and other slick Surface Web sites, customers were asked to rate Silk Road sellers and post feedback about mundane matters of shipping and packaging. And similar to the way Amazon offers the opportunity to leave long reviews of goods purchased on the site, Silk Road buyers posted descriptive comments of their drug highs and lows—and gave lots of friendly drug-loving advice.

Like any good buccaneer, Dread Pirate Roberts had a code: No stolen items or child abuse material could be sold on Silk Road. This added to the impression that the site was actually a force for moral good. "Silk Road is about something much bigger than thumbing your nose at the man and getting your drugs anyway," Ulbricht wrote. "It's about taking back our liberty and our dignity and demanding justice."

For the two and a half years that Ulbricht worked anonymously at the controls, Silk Road attracted a base of several thousand sellers and more than one hundred thousand buyers and was said to have generated more than $1.2 billion in sales revenue. According to a study published in *Addiction,* 18 percent of drug consumers in the United States between 2011 and 2013 used narcotics bought on the site.

Court records show that federal agents used traditional, time-honored investigatory tricks to pursue Ulbricht, along with some cutting-edge, classified, high-tech cyber-sleuthing, and were eventually able to dismantle Silk Road. For any black marketer, staying "dark" all the time is extremely difficult. All it can take is one small mistake to expose your real identity. Due to some missteps online, or weak computer code, Ulbricht didn't sufficiently cover his own tracks—which led agents to locate him and link him to the website. This famous case shows that with enough resources, law enforcement can unmask those on Darknets. By the time the cyber-moralist was arrested in the science-fiction section of a San Francisco public library—his fingers on the keyboard and logged on to the Silk Road site using the free Wi-Fi—the FBI estimated that his digital black market had brought him $420 million in commissions, making him, as *Rolling Stone* put it, "one of the most successful entrepreneurs of the dot-com age."

What helped Dread Pirate Roberts become rich beyond his wildest hippie dreams? Cyber effects and the cyber environment itself. As any big retailer knows, the best way to make a killing is to out-psych the shopper—and manipulating the shopping environment and shopping experience is key. This is why the air temperature inside large grocery store chains is often uncomfortably chilly and why distractingly loud music plays in clothing stores: Being cold, overstimulated, or confused can provoke impulsive shopping.

Now imagine how the cyber environment facilitates the online shopper of illegal merchandise. Anonymity and invisibility online would certainly encourage shopping, particularly for people who previously had to hide from authorities. If you add online disinhibition to this mix, you find an emboldened consumer who is even more undeterred by risk. Next, let's factor in what I described as the cyber-risky-shift phenomenon in the chapter about teens online. If a large number of people online are doing something, this can make risky or extreme behavior seem normal.

The final ingredient is greed. Think about how opportunism factors into this equation, and how tax-free financial gains facilitated by the amassing of anonymous and untraceable wealth would encourage more vendors to participate and offer a greater selection of illegal goods. Having more vendors creates more opportunities for drug buying. More sales means a greater demand for supplies, whether it is cannabis or poppies or pharmaceuticals. The result?

A booming cybercriminal economy. It's a virtual bubble. Silk Road's "off the grid" computer servers were hidden around the world, in places like Latvia and Romania, but after these were uncovered by a team of agents from the FBI, the DEA, the IRS, and U.S. Customs, and the illegal transactions monitored, law enforcement was led to suspected drug dealers and buyers in the United States, Britain, Ireland, Australia, and Sweden who were using Silk Road. "These arrests send a clear message to criminals," said Keith Bristow, director of Britain's National Crime Agency, after the arrest of four men for alleged drug offenses. "The hidden Internet isn't hidden, and your anonymous activity isn't anonymous. We know where you are, what you are doing, and we will catch you." Then Bristow nicely added the good news: Criminals, he said, "always make mistakes."

According to U.S. District Judge Katherine B. Forrest, who sentenced Ulbricht at his 2014 trial, Silk Road created drug users and expanded the market, increasing demands in places where poppies are grown for heroin manufacture. The black market site had impacted the global market. The prosecutors alleged that, in addition to becoming the Jeff Bezos of drugs and guns, Ulbricht had ordered up and paid for the executions of five Silk Road sellers who had threatened to blackmail him or reveal his identity. (Nobody was actually killed; it was part of the sting operation.)

He was found guilty of seven drug and conspiracy charges and was given two life sentences, another for twenty years, another for fifteen years, and another for five years, without the chance of parole. The Bitcoins he left behind on Silk Road's servers—worth an estimated $180 million—were forfeited, and auctioned off by federal marshals starting with an installment of fifty thousand Bitcoins.

Ulbricht pleaded for leniency. His parents did too, testifying that their son was not a hardened pathological criminal and, if freed, presented no more potential harm to society than any other former lawbreaker. He was just a conscientious barefooted free spirit whose ideals of personal and economic freedom spiraled out of control.

The judge did not show leniency, however. In her sentencing, Forrest told Ulbricht that what he did "was terribly destructive to our social fabric." Prosecutors traced the deaths of six people who overdosed on drugs back to Silk Road, and two parents who lost sons spoke in court. One victim, an athletic young man identified only as "Bryan," was an employee of a small money-management firm in Boston. He had struggled to resist drugs, and seemed to be winning the battle. He died of an overdose from heroin purchased on Silk Road just days prior to Ulbricht's arrest.

Bryan's grieving father's statement in court conceded that his son had made bad choices in life, but felt that the operation of Silk Road had made his struggle against drugs even harder, as it had "eliminated every obstacle that would keep serious drugs away from anyone who was tempted."

Lawlessness in a lawless environment required that an example be made. But in the case of Dread Pirate Roberts, at the age of thirty-one, he wasn't forced to walk the plank. He was given life.

The Pirate Bay

Criminologists will tell you that one of the first entry points into criminal activity is ambivalence about the law—and a lack of regard for the rules of society, particularly if these rules are keeping you from an activity that's either enjoyable or economically rewarding. In other words, you don't have to descend to the Darknet underworld to find opportunities to get caught up in criminal activity online. It can begin benignly.

As the *Washington Post* put it on December 10, 2014—the day after the Swedish government finally shut down the world's largest pirating site:

> In the late hours of Tuesday night, the Pirate Bay abruptly disappeared from the Internet, the result of a surprise raid on the site's servers by Swedish police in Stockholm.

But forget the big-picture questions of Internet freedom or intellectual property. The real problem, for millions of Internet users, is *How am I going to watch TV?*

If, like me, you aren't into shoplifting and you haven't joined the tech-trendy anti–intellectual property movement, you might need a few things explained. Software piracy is the mislicensing, unauthorized reproduction, and illegal distribution of software for business or personal use. The Pirate Bay was a site where, once you had the proper software, you could download large files for no charge—free games, videos, movies, songs, books, or apps. The stated intention of the Swedish activists who founded the site in 2003 wasn't content theft, criminality, or bald opportunism. Actually, they said they weren't in it for the money. Like Ross Ulbricht, they claimed it was an effort to disrupt the corrupt capitalist marketplace.

There it is again: a common code.

The anti–intellectual property movement is popular with the tech industry—and has lots of fans in cyberspace. Shrewdly, the administrators of the Pirate Bay publicized their site by organizing political rallies and circulating petitions about "the practical, moral, and philosophical issues of file sharing." It claimed it was all about mutuality and reciprocity and brotherhood, not stealing.

In cyberspace, "torrent" refers to any type of file that is shared via the protocol BitTorrent, a peer-to-peer system (in tech-lingo, "p2p") for sharing big files. Rather than coming from just one server, segments are taken from multiple sources at the same time. The result? Lower bandwidth, so delivery is faster. This makes the pirating easy and quick, something that is mandatory for success online.

It's a little like fast food: probably best not to think about quality, nutritional value, or place of origin. But at least when you bite into a Big Mac, you have the moral comfort of knowing you paid for it.

Napster was the groundbreaking site that pioneered mass music file-sharing and not-paying. In just two years, between 1999 and 2001, it acquired 60 million registered users until the Recording Industry Association of America (RIAA) took the company to court and forced it into liquidation. Law enforcement caught up with the Pirate Bay eventually too, and its founders were charged with copyright infringement. But the site stayed up, often shifting to new domains until it settled on a secure, cloud-based infrastructure in 2012. By then, it had tens of millions of Internet users.

How big was the impact of peer-to-peer file-sharing? The recording industry, already grappling with challenges from new technologies in the shift to digital, claims there was a 45 percent drop in music sales in the first decade after Napster was launched. In the first five years that the Pirate Bay was in operation, approximately 30 billion songs were illegally downloaded on file-sharing networks. These financial losses compound once you consider the millions of lost jobs, failed businesses, and uncollected tax revenue—and that's before you factor in the impact on undiscovered creative talent.

But there were other costs, possibly greater. Once it began, the trend of pirating continued to become more popular, more prevalent—and cool. In 2010, an episode of *Game of Thrones* broke a piracy world record when it was shared 1.5 million times on the Pirate Bay within the first twelve hours of its original airing.

Rather than throwing a colossal and very uncool tantrum, Jeff Bewkes, the CEO of Time Warner, took another approach. "*Game of Thrones* is the most pirated show in the world," he bragged. "Well, you know, that's better than an Emmy."

Hmmm. Is it really?

It costs HBO about $6 million to make each episode of *Game of Thrones,* two or three times the budget of a typical network or cable show. Could the executives at the premium cable channel really have been happy about that much lost revenue?

Or does HBO have to play along, just to stay popular? I guess that's the cost of being cool. Which brings me to another puzzling conundrum of human behavior: Why do people break the law when they know it's wrong?

Justifications for content theft online are plentiful and apparently compelling, according to the many studies that have been done on the behavior. To start with, some do not regard this as theft at all but more like sharing a library book. Next, there's also considerable difference in how people regard the theft of tangible property versus intangible property, even though, from a moral and ethical perspective, there is no difference.

What do I mean? A person who feels nonchalant or ambivalent about downloading pirated material probably would never steal a CD from a record store or a yoga mat from a yoga studio. Tangible theft is commonly regarded as an ethical breach—a crime. But intangible theft seems . . . well, *less real.*

Everything is intangible in cyberspace. Does this mean that people are more willing to steal there? Some studies argue yes.

Another explanation for the prevalence of online theft (and other crime) may relate to the cyberpsychology construct *minimization of status and authority online,* which I flagged in a 2014 Europol report. There is little apparent governance in cyberspace, and therefore the risks of getting caught are perceived as being minimal. Why pay for an HBO service if there's practically no negative repercussion?

The propensity to engage in online theft could also be fueled by the effects of perceived anonymity and online disinhibition. Last, the prevalence of online theft itself creates a shift in social outlook. When so many people are talking openly and posting about having pirated the latest *Game of Thrones,* it can change the perception of wrongdoing. It can normalize and socialize the activity. Sooner or later the law can cease to reflect the moral perceptions of the crowd. That's the funny thing about moral perceptions: If enough people are breaking a law, it stops being regarded as immoral. As in the case of extreme cyberbullying and other problematic behaviors described in previous chapters, the fact that individuals who are inclined toward online piracy can easily syndicate with large groups of like-minded people means they are much more likely to regard their own behavior as normal.

This kind of cyber-socialization occurs each time an individual visits a pirating site where the index shows uploaded content from thousands of participants around the world.

How could so many generous file-sharers be wrong?

Attempts at deterring the behavior have also been ineffective. People have sought to make examples of piracy with widely publicized arrests, like the fifteen-year-old boy in Sweden who used his school servers to illegally download twenty-four movies, including *The Social Network, The Mechanic,* and *The Fighter.* Soon afterward, a virus was discovered on the school servers, which was traced to him. Rather than simply calling the boy into the principal's office for a tough chat, and assigning punishment for both introducing the virus and downloading illegal files, the school policy required that a call to the police be made for any actual crimes on campus. While more than eighteen thousand people were sued by the recording industry for illegally sharing music in the mid-2000s, most of those cases were settled out of court. The case of the fifteen-year-old boy in Sweden, though, was not. He faced up to two years in jail.

Does that seem fair? Probably not to the boy in Sweden. And probably not to millions of others who argue that piracy is a victimless act that causes no real harm to artists and the recording industry, both of which could afford the financial loss. It doesn't seem to matter to them that it is still stealing. I enjoy the discussion of ethics and morals online, but it's troubling to me that nobody else seems to care. What it says to me: A new norm has effectively been created.

In real life, a new norm can be created quickly, in a generation. But as I have discussed in previous chapters, due to cyber effects, norms evolve online at the speed of light. How long did it take before we were all taking selfies—or before the *Oxford English Dictionary* had anointed the word? Changes occur so quickly online that it is hard to keep up. We may be raising a generation of what I call *virtual shoplifters* who have a different sense of beliefs about property rights, privacy, national security, and authority.

Does that sound cool?

Coolness is just another aspect of peer pressure that serves the creation of new norms. Coolness is a way to win approval for cleverness, for knowingness, and for being an early adopter of new technology, new behaviors, and apparently new ethics. Coolness can also win you a free one-way slide down the moral slippery slope.

The cyber environment, whether we're talking about the Surface Web or the Deep Web, makes the slope even steeper and slipperier. Even with the popularity of streaming sites like Spotify and Pandora, which offer better sound quality and custom channels—a business model that was essentially designed to combat piracy—the robbing of songs and movies and TV shows goes on.

A move to decriminalize piracy in the U.K. is a further indication that society's norms, impacted by technology, are beginning to dictate the law. After four years of discussion, and despite pressure from the entertainment industry—which asked for Internet service providers (ISPs) to keep a database of suspected illegal downloaders and to threaten prosecution—the government has decided to send persistent file-sharers "educational" warning letters. A maximum of four letters, which will escalate in severity, will be sent. But there will be no threats of legal action. Why? "We found that many people are not necessarily aware that what they are doing is illegal," said a spokesperson for the U.K. Department for Business, Innovation and Skills. "The new alerts make users aware of the impact of illegal downloading and will help promote the use of legal digital content." This amounts to little more than a slap on the wrist.

In other words, on the high seas of cyberspace, it appears the pirates have won.

Psychology of the Hacker

Not long ago, I was invited to join the Steed Symposium panel on cyber-security at the Los Angeles Film Festival and found myself sitting on a stage with blinding lights, awkwardly perched on a very tall director's chair. I don't want to go on too long about this chair, but it was really cool-looking and probably the most uncomfortable thing ever to sit on—a narrow seat, rigid back, unnaturally and uncomfortably high off the ground, with a tiny band of wood for a footrest, only an inch wide, which seemed to be the only thing keeping me from tumbling down to the floor in front of several hundred people.

Next to me, perched on his own precarious chair, was a famous ethical hacker, Ralph Echemendia, a brilliant self-taught tech expert who had recently served as the subject-matter expert on *Snowden,* the Oliver Stone movie.

Ralph was radiating swagger, and had a unique, ineffable vibe—like a cross between a Mexican gunslinger and the lead singer in a heavy metal band. Based on our differences—I am definitely not cool, am not self-taught, and am pro-governance—and based on what I knew about hacker culture, I suspected that Ralph and I would have nothing in common and very little to say to each other.

I looked out into the audience. Their faces were like those of the participants of all the other cyber-security conferences I've attended—they had that look of grim determination and *endurance*. The same issues and problems would probably be discussed. And no one was likely to come away any wiser. And meanwhile, it wasn't going to be much fun, either.

At cyber-security conferences and summits, while the participants are talking about vulnerability or privacy for the one-millionth time, I am always thinking, *So what is the motive behind this behavior? What is the cyberpsychology?* One way or another, everything that involves human beings must come back to human behavior—and to motives—no matter where they are, including cyberspace. Which brings me to an interesting question.

What is the motive behind hacking?

The common definition of a *hacker* is a person who secretly gets into a computer system intending to do damage. Since the earliest days of computer networks, the most highly skilled technicians, programmers, and coders at the world's foremost university computer-science departments have enjoyed swapping stories of hacking, sometimes done as a practical joke, sometimes done more maliciously. There are lots of examples of how a clever hacker has exploited social convention to gain entry by using private information, such as finding out someone's birthday and sending them an email with a masked and malicious link. This kind of socio-technical approach runs through the whole phenomenon of malware and other cybercrime tools such as worms, Trojans, spyware, keyloggers, ransomware, and rootkits, to name a few. Because hackers understand both tech and human behavior, they have an advantage over those who don't, which only makes a stronger case for all of us becoming more savvy about human behavior online, in order to properly protect ourselves.

As with pirates, there are folkloric aspects to hackerdom, as well as a David-and-Goliath story line that appeals to young people. Hackers seem magical, almost superhuman. And like superheroes, they are often individuals who are regarded as weak-bodied nerds in the real world. That's until they deploy their special powers—the brains and the tech skills to upset an entire corporation, a bank, a health insurance provider, or an entire nation. Like Superman's nemesis, Lex Luthor, some claim they can bring down a power grid and turn the lights off in a vast metropolis with a click of their index finger. They are swashbuckling in their way, inventive and courageous, as well as defiant.

And, yes, pretty smart.

Motives for hacking? There's an array of them: boredom, emotional release, monetary gain, curiosity, political or religious or philosophical beliefs, sexual impulses. Before we get to that last one, let's look at hacker culture.

Hackers also have a famous code, or something like a code, "The Hacker's Manifesto," an essay written by "The Mentor" (a.k.a. Loyd Blankenship) that has been shared for three decades now, since 1986—and still has relevance today:

> Another one got caught today, it's all over the papers. "Teenager Arrested in Computer Crime Scandal," "Hacker Arrested after Bank Tampering." . . . Damn kids. They're all alike. But did you, in your three-piece psychology and 1950's technobrain, ever take a look behind the eyes of the hacker? Did you ever wonder what made him tick, what forces shaped him, what may have molded him? I am a hacker, enter my world. . . . Mine is a world that begins with

school. . . . I'm smarter than most of the other kids, this crap they teach us bores me. . . . Damn underachiever. They're all alike.

Talented and lawless, beyond society, the criminal hacker's story could be taken from the pirate playbook, like a mash-up of *Bluebeard* and *Revenge of the Nerds* with a dash of teenage narcissism à la Holden Caulfield thrown in. And while Edward Snowden isn't truly one of them, his sharing of confidential NSA files raises the same questions that are raised by the persistence of hacking in our culture.

Is it heroic or criminal?

Are hackers courageous—or just angry?

According to technology writer Debra Littlejohn Shinder, typical criminal hackers share a set of personality traits:

- They have a tolerance for risk.
- They tend to have a "control freak" nature and enjoy manipulating or outsmarting others.
- They have advanced tech skills (to varying degrees) but at the very least can manipulate code.
- They have a certain disregard for the law or rationalizations about why particular laws are invalid or should not apply to them.

The media and popular entertainment often use the term *hacking* in a derogatory way. A more accurate view of hacking is not always negative. While there are still hackers, or "black hats," who violate cybersecurity due to pure maliciousness or for personal gain, like stealing credit card numbers (or "carding") or cooking up a new virus, there has been a rise in "white hats" or "ethical hackers," who use their skills for good, ethical, and legal purposes. They are often employed by large organizations to test their computer security systems. This is called *penetration testing*. If these hackers find vulnerabilities, they will disclose it to their client.

Falling somewhere in between are the "gray hats." They don't work for their own personal gain but may do unethical things or commit crimes, which they tend to justify as for a good cause. For instance, they might test the security of a cyber system, looking for vulnerability, and let the operator know about it.

Suppose, say, that you woke up in the morning and heard a knock at your bedroom door. You opened it and found an ethical burglar standing there, wanting to tell you that he had broken into your house, bypassed your alarm system, entered your bedroom the night before, and left a note by your bedside. Your house has a serious security flaw. What would you say to this intruder? "Thanks"? That's what gray-hat hackers do, and they don't always get thanked for it.

At the start of the Steed Symposium, after introductions were made, a short film was aired. It was set in the future—2024—and told the story of a woman who had been given a brain implant, a chip that regulated and controlled her. She had committed a murder, in fact. But she had no motive. She had been directed to kill a man by the chip implanted in her brain.

The audience perked up—hey, maybe this was going to be a good night after all.

The moderator turned to Ralph Echemendia, the cyber-security samurai, and asked how he first got into hacking.

"I was a thirteen-year-old boy growing up in South America, and my friends and I were getting into porn," he said. "*And it never downloaded fast enough!* That's how I got into hacking."

The audience loved that—and burst into laughter and applause. Ralph continued talking about his early years—how he hacked ham radios, hacked old bulletin board systems, and did phone phreaking, or finding ways to mess with the telephone company, usually to get phone service for free. His interest in technology eventually led to jobs in the computer industry. For the past fourteen years, he has conducted security audits and penetration tests, and has consulted for numerous organizations around the world, including the United Nations, Oracle, and various hospitals and financial institutions.

As I listened to Ralph talk, his passion for his work shone through—and began to shatter my narrow assumptions about hackers and hacker culture. It became increasingly apparent that intellectually, in terms of our view of all things cyber, we were aligned. He didn't talk about tech as much as he talked about people's lives, about culture and society. And the ways that technology could be used to improve life on earth, not distort it.

The moderator turned to me next. "What's the explanation for why people hack?" she asked.

"If you are talking about humanistic psychology, it could be for an emotion such as love or revenge," I replied. "But if we are talking behaviorism, then it's all about reward or profit. But my favorite explanation for why people hack is the Freudian, or psychoanalytic, school of thought."

The moderator looked puzzled. Ralph looked intrigued.

"Psychoanalytic? What's that?" she asked.

"It explains hacking as a psychosexual urge to penetrate."

"Respect!" Ralph cried out, and fist-bumped me. Our friendship was born.

This is a line I've used before—primarily to wake up an audience of near-comatose cyber-security professionals. It is meant to be a joke aimed at the behavioral sciences, where there are typically several conflicting explanations for one phenomenon, which can be so irritating to the dyed-in-the-wool hard-science community. Recently, though, I was unmercifully trolled on Reddit by technophiles who felt offended by this joke.

I read through the stream of abusive comments, and to be honest, I was pretty impressed by the level of psychoanalytic knowledge expounded, everything from my "father complex" to my alleged desire to—how can I put it delicately?—*be intimate with* a hacker. When some commenters actively defended me, I resisted the urge to jump into the conversation and thank them. All in all I was not outraged or shocked or hurt. I saw this trolling behavior for what it was—simply interesting feedback, data, and lots of it.

As my good friend John Suler says, "Let your critics be your gurus. You can treat them as an opportunity. Ask yourself why you're ruminating on a comment. Why does it bother you? What insecurities are being activated in you?"

In other words, nobody can make you feel anything. You are responsible for how you interpret, react, and feel. It's good advice to keep in mind when dealing with barbs and nasty comments online. If you are hanging out in cyberspace, you will surely find them.

Back in the real world, later that week, Ralph and I had dinner. We talked for hours. We discussed everything from the cyberpsychological nuances of socially engineered attacks to how easily your mobile can be compromised to send mischievous texts. And we discovered we shared a passion: kids with tech skills and how to nurture their talent. Like another colleague of mine, FBI Special Agent in Charge Robert Clark, a superdedicated and charismatic man who is very concerned about keeping young teens out of trouble in cyberspace and the real world, Ralph and I have both seen the statistics showing that younger and younger kids are becoming involved in hacking—and crime online.

Surely the generation being raised will have unimaginably fine tech intelligence. We've spent decades rewarding individuals with a high IQ, and more recently EQ (emotional quotient). But what about a new metric, TQ—technology quotient—to identify, assess, acknowledge, and reward individuals with the superlative tech skill sets that many kids intuitively display? Is a metric for intelligence designed almost fifty years before the first computer and one hundred years before the ubiquity of the Internet still fit for the purpose?

We need to find ways to reach out to tech-talented individuals, especially young people, nurture them, and teach them to think about others as people—not computers or machines. The tech-talented have such a lot to offer. And just like the pirates of yore, sailors who could turn a frigate on a sixpence and navigate expertly by the stars—and with the right environment and nurturing, could have made great naval commanders—the skills of high-TQ individuals could be harnessed to make enormous contributions to the quality of all our lives, or cyber-lives.

As Ralph spoke, I was beginning to see that hackers have their own distinct perspective and moral code. And while I certainly don't endorse anything that involves breaking the law, I do respect raw talent and genius. And if hacker culture can produce a guy like Ralph, there must be good things happening there.

At the end of dinner, Ralph said, "Mary, the way you understand behavior online, you have mad hacking skills."

What did he mean?

"But, Ralph, I'm not a hacker."

"Oh, but you are—you just don't know it."

Crypto-Markets

After the arrest of Ross Ulbricht and the shutting down of Silk Road in 2013, it wasn't too long before a new site, Silk Road 2.0, sprang up to fill the void. There were lots of copycat sites on the Darknet selling contraband by then—sites like Evolution, Agora, Sheep, BlackMarket Reloaded, AlphaBay, and Nucleus—often referred to as *crypto-markets* by law enforcement.

Many of these have come and gone already, but the offerings continue to expand. The black market has proven amazingly resilient. And the sellers grow more sophisticated each year.

As an article from *Wired UK* attests:

> The first thing that strikes you on signing up to Silk Road 2.0 is the choice. There were almost 900 vendors to choose from, selling more drugs than I'd thought possible. Heroin, opium, cocaine, acid, prescription drugs are all readily available. Technically speaking, Silk Road 2.0 is an anonymous market for anything (with some exceptions, such as child pornography), which means there are also sections for alcohol, art, counterfeit, even books. Listings included a complete boxset of *The Sopranos;* a hundred-dollar Marine Depot Aquarium Supplies voucher, and fake UK birth certificates. Each with a product description, photograph and price.
>
> But most people are here for the drugs. . . . As I browsed through the marijuana offers, I found 3,000 different options advertised by over 200 different vendors.

According to some accounts, the number of products available on Darknets had more than doubled in less than two years after the 2013 arrest of Ulbricht, to fifty thousand.

Why?

I suspect the swashbuckling stories in the media about Silk Road may have encouraged curiosity about the Darknet and its offerings. The profusion of how-to guides that help newbies and first-timers figure out how to get to Darknets is also a factor. According to INTERPOL, as of August 2014 there were at least thirty-nine such markets, and the majority use English, although there are sites in French, Polish, and Russian too. An investigation in 2013 estimated that one-quarter of the illegal substances sold in the U.K. were obtained from them. We can't know for certain, but the percentage of drug buyers using Darknets in the United States could be as high, or higher. A study done in 2015 analyzing the size of Darknet markets found that they do a brisk business. In just four years, since the development of the original Silk Road, the total sales volume is generally stable, around $300,000 to $500,000 a day. Even more remarkable, anonymous marketplaces have proven to be resilient to takedowns and scams, because demand plays a dominant role.

What does that tell me? If we believe that figure—that as much as one-quarter of the illegal drugs in the U.K. and U.S. are obtained through Darknets—then it means one-quarter of those drug buyers have taken the step to download the suitable protocols like Tor and have learned how to use them.

And it means that one-quarter of these drug buyers have arranged for shipping of illicit goods to their residences or post-office boxes. The United Nations Office on Drugs and Crime (UNODC) review of global drug seizure data shows that cannabis seizures obtained through the postal service rose 300 percent in the decade from 2000 to 2011.

It means that one-quarter of these buyers are likely exchanging cryptocurrency—or using some form of anonymous and untraceable method of payment.

In 2015 the UNODC confirmed that there had been no major change in the regions where illicit crop cultivation and drug manufacture take place:

> . . . but the illicit drug markets and the routes along which drugs are smuggled continue to be in a state of flux. The "dark net," the anonymous online marketplace used for the illegal sale of a wide range of products, including drugs, is a prime example of the con-

stantly changing situation, and it has profound implications for both law enforcement and drug trafficking.

We know from reliable field reports and investigative journalism that teens in particular have flocked online to buy drugs in recent years. It is perceived as being safer than entering a bad neighborhood. They may be looking for a quick way to score pot, ecstasy, or some other party drug. They may be using these drugs themselves—or selling them for a profit to friends. Or, like the pirates of old, they may be simply looking for some excitement and adventure.

Now let's consider what we know about this age group. We know that impaired judgment can be common in teenagers, and when they gather in groups, due to the effects of the risky-shift phenomenon, they are even more likely to be judgment-impaired. Their judgment is further compromised in cyberspace due to the effects of online disinhibition.

Now let's put these factors together with the act of buying drugs, now made as easy and prevalent as pirating music, and ask a new set of questions: Would a teenager be more likely to try a new drug when anonymously browsing the thousands of offerings on a black market site, simply due to the vast selection—so temptingly described and photographed—than he would to buy the same drug on the street?

Probably.

Would a young person be more likely to buy more drugs, due to the effects of online disinhibition?

Probably.

Remember the Triple A Engine of the Internet from chapter 2? The three ingredients—affordability, accessibility, and anonymity—are known to successfully drive people to sites that facilitate sexual communication online. But I believe this construct also explains the success of the black market drug sites. In other words, if you offer something illicit and forbidden, but with the features of the Triple A Engine, buyers will appear in droves.

Okay, then, does this mean that an individual risks a higher chance of becoming involved with drugs due to all the aspects of the digital marketplace?

In my opinion, yes.

There were roughly 187,100 drug-related deaths worldwide in 2013, according to the UNODC, and some 27 million people worldwide had problems with drug use. Obviously, this is the biggest downside to easy illegal drug sales.

As tragic as those numbers are, there is another potential downside to the prevalence of drugs online that could have even greater ramifications. Buying drugs online, particularly in Darknets, requires an individual to enter a "neighborhood" where bad things can happen. There's actually a well-known criminology construct to explain it.

Cyber-RAT

A famous Canadian criminologist, Kim Rossmo, found that great white sharks and serial killers have common behavioral traits, in terms of how they hunt their prey. They both are focused killers, have a strategy, prefer their victims to be young and alone, and like to attack when light is low.

It would be interesting to consider the hunting patterns of cybercriminals in a place that's also dark, and where there are plenty of young victims, surfing all alone.

A number of interesting things can happen when a young individual enters a new society and culture, such as the ones that exist on the Darknet. To begin with, consider how the good manners and responsiveness of vendors may send distorted social cues. Like what?

This is a safe place.

This is a place where people are looking after you.

This is a place where they really care about your business—and getting that five-star good-vendor rating.

And this is a place where you could make cool new friends.

In an age when most young people spend so much of their time online, where they make and maintain their social contacts, wouldn't they wind up making some new acquaintances on Darknets?

In my own work investigating the evolution and behavior of the cybercriminal, I have been influenced by the pioneering work of David Canter, an investigative and environmental psychologist in the U.K. whose great book *Mapping Murder* is a fascinating read. Canter's main areas of work are real-world offender profiling and geographical profiling. These theories can be used to demonstrate how environment can impact criminal behavior, and have helped me consider the impact of the cyber environment on crime.

As Canter states: "Criminals reveal who they are and where they live not just from how they commit their crimes, but also from the locations they choose." In my work, I consider how the cyber location reveals the criminal.

We actually know a lot about pathways into a life of real-world crime due to the abundance of academic work in this area. And I do mean *abundance*. In the field of criminology, there are biological theories, labeling theories, geographical theories, trait theories, learning theories, psychoanalytic theories, addiction theories, and arousal theories. But if you wanted to know how, specifically, a young person goes from curiosity about the Darknets to cybercrime, or being part of organized cybercrime, we are still putting those pieces together.

As an advisor to Europol, I am currently one of the principal investigators on a new research initiative that will look at how young people get drawn into cybercrime—and specifically what the pathway is from cyber juvenile delinquency to lone cybercrime to organized cybercrime. One of the established criminology theories that we will be experimenting with—and trying to apply to cyberspace—is *routine activity theory,* or RAT.

Many theories focus on the individual characteristics of criminal offenders, but RAT, which was first introduced by sociologists Lawrence Cohen and Marcus Felson in 1979, examines the environments where crimes occur. The theory maintains that *when motivated offenders and suitable targets meet in the absence of capable guardians, crime is likely to happen* (motivated offender + suitable target + absence of capable guardians = more crime).

What's helpful about RAT is that the absence of any of these three conditions can be enough to prevent a crime from happening.

The theory is based on human nature—and the patterns of everyday life that all of us fall into. As one criminological handbook puts it: "Individuals have different routines of life—traveling to and from work, going to school or attending religious functions, shopping, re-creating, communicating via various electronic technologies, etc.—and these variations determine the likelihood of when and where a crime will be committed and who or what is the victim."

Criminals have patterns too—places where they live, work, and play. We know, for instance, that if you enter a real-world neighborhood where more criminals live, you are more likely to be a victim of crime. Now, what if that neighborhood is not policed in any effective way? Crime goes up. Your chances of being a victim go up too.

Now let's consider this in terms of RAT—or what we might as well call cyber-RAT.

The criminal neighborhoods online—on the Deep Web. How many motivated offenders are there?

Hundreds of thousands.

Suitable targets?

Even more.

How about capable guardians?

You know the answer already.

In the real world, young people have friends, older siblings, parents, neighbors, shopkeepers, teachers, and police who will say, "Don't stand on the table!" or "Don't run with scissors!" or "Don't walk near that ledge!" or "Don't go to that neighborhood!"

But in cyberspace, authority is minimal and there is a perception that nobody is in charge.

Because nobody is.

Now let's consider this new bad cyberspace neighborhood, the sites on the Darknet. Imagine a boy who grows up in poverty, and in a real-world bad neighborhood that is populated by gangs. Very likely, that boy will grow up with lots of insights into criminal behavior—and have honed instincts or antennae about criminals. He'll have "street smarts." And because of this, he'll have a pretty good understanding of gang culture and know the protocols and the rules. (Because there will always be rules.)

One of the rules of a gang is that once you are a member, you are a member for life. A boy who grows up in a gang neighborhood will know this, almost instinctively, due to the experiences and environment of his childhood.

Now let's think about a boy raised in a suburb of Tallinn, Estonia. He may be socially isolated, spending a lot of time online in the safety of his bedroom. He has superb tech skills but no street smarts. He knows nothing about gang life and gang culture. He will have no experience, no wisdom. And yet it will be as easy for him to wander into Darknets as any boy. The boy in suburban Estonia can virtually transport himself within minutes into a high-crime neighborhood, where his tech skills will be viewed as a commodity. And there he could be groomed, coerced, or tapped to join a community that is really a cybercrime gang.

Once he falls into this cybergang, can he get out?

There are hundreds of stories of kids such as I've described, and we know this because of the number of mules, or money launderers, that are used by cybercriminals. Sometimes they answer an online ad—or post on a university bulletin board—that offers an opportunity to "stay at home and earn money." The job is described as a financial manager, an overseas representative, or a payment processor for a new online business. *No experience needed.* Even some well-known businesses have had their brands hijacked and misused in this way. The job entails receiving money from customers, deducting a commission, and then wiring the balance overseas, usually to a bank in Russia or Eastern Europe. The offer often seems almost too good to be true.

The term *mule* comes from the drug-smuggling trade. A mule is paid to transport drugs for a fee. A *money mule,* or *smurfer,* transports money, not drugs, acting as a money-laundering service. Because this kind of illegal action doesn't involve tangible illegal goods, it may seem less risky, and many are not aware they are doing anything illegal until the police knock on their door. When the crime operation goes down, the mules are often the only ones who get caught and punished. By then, the money is in the hands of the cybercrime ring.

Where does this money come from? Usually stolen credit cards. In the past decade, banks worldwide have lost billions of dollars to cybercriminals. Quantifying the amount of money, and the actual threat, has proven difficult. But one thing appears certain: The incidence of teenagers turning to cybercrime is on the rise.

Cybercrime Rings

The world was shocked when Crimebook, one of the world's largest English-language crime forums—which offered advice about how to wire funds anonymously, purchase illegal items online, and pay for other illegal services—turned out to be founded and operated by three teenagers in the U.K. In 2011, their forum had 8,000 members worldwide and was linked to a trading site, Gh0stMarket, that had embezzled £16.2 million by hacking into 65,000 bank accounts. When the teens were arrested and their computers seized, the details of 100,000 stolen credit cards were found.

Just do a search on the words *teenagers* and *cybercrime* and you'll see enough examples to convince you of an epidemic. In a survey done by Tufin, an online security company, roughly one in six teenagers in the U.S., and one in four teenagers in the U.K., were found to have tried some form of Internet hacking.

The explanation for this is demand. According to a recent Europol report, cybercrime has evolved from a few small groups of hackers into a thriving criminal industry that costs global economies between an estimated $300 billion and $1 trillion a year.

What does demand have to do with that? The Internet is where criminals buy and sell products and services, as well as hire and train. The underground economy relies on websites and forums for recruiting. As criminals network, they share experience and expertise, and meet cohorts. On various sites, there are forums to learn, step-by-step, the best ways to commit identity fraud, to steal credit card information, or to convince an elderly person to wire you, a perfect stranger, money overseas. There are even tutorials.

As the online criminal economy grows, it needs new accomplices and new victims. Frequently it's difficult to know which is which.

It's easy to see why a young person might believe that buying drugs in this environment is safer. As all the studies of piracy have shown, the first thing a person worries about when purchasing anything illegal is the risk of being caught by authorities—whether it's their parents or the police. And if you feel safe somewhere, you are likely to spend more time there, whether you are buying drugs, looking at new sites, or just hanging out in a forum.

But it isn't safe. Effectively, these young people are entering a high-crime environment that's largely unpoliced (except by other criminals), where scams are rampant, and where many illicit things besides party drugs are available.

Not to mention it's where lots of nasty things can happen to your computer and personal data. By visiting a pirating site, you can become vulnerable to more than just viruses and malware. Because of the nature of file-sharing being peer-to-peer and decentralized, each user acts as a server for others. Many file-sharing sites recommend that users stick with program defaults, thereby making their folder shareable—and their device can potentially wind up becoming invisible storage sites for anything a cybercriminal needs to hide and find later, quickly, from lists of illegal credit card information or thousands of child abuse images. When anybody uses a p2p service site, they can unwittingly download encrypted files containing illegal content. This is called being a "storage mule."

According to Adrian Leppard, the Commissioner of the City of London Police, a quarter of organized crime in Britain involves online fraud, bringing in tens of billions of pounds in profit a year. "When many of the offenders are abroad and they are using the Internet, which is unregulated, it is very difficult to see how a traditional enforcement approach will solve the problem," Leppard said in an interview with the *Telegraph*. "Even if we had ten times the number of police officers, I am not sure that would necessarily address the problem."

Cybercrime isn't just escalating, though.

It has changed criminal culture.

Ubiquitous Victimology

We have always had an underworld, whether in mythology or in real life. In timeless fables, heroes are often forced to descend into underworlds and a psychic conflict ensues. The battles with demons and monsters are metaphors for the hero's own inner struggles and conflicts. In psychology, it was Sigmund Freud who created the concept of the human unconscious as a dark realm that required special knowledge to access.

Freud believed that the antisocial elements inside us seek a dark place to hide, much the way underworld criminals look for covert environments—and will always find them. There will always be crime. There will always be criminals.

My question: *But do we have to make it so easy?*

Whether the descent to the Darknets is to buy a stolen credit card number or pull off a data breach of the IRS, it has never been simpler or easier to become a criminal. It used to require physical risk. It meant armed holdups, face-to-face encounters with real guns, and a need for outsmarting law enforcement. Now, for most cybercriminals, the risks are negligible. Only the poor mules—or teenagers from the Estonian suburbs—get caught.

As for victims, they're everywhere. Once upon a time, crime was visited upon you if you walked alone at night or walked in the wrong neighborhood, where your wallet or car keys could be stolen. Or it happened randomly. You were in the wrong place at the wrong time.

Now everywhere you go, it's the wrong place and time. With many of the greatest treasures you possess—your identity, your data, your account numbers and passcodes, your passport information, and the rest—now kept in cloud storage or available via your smartphone, the opportunities for theft are endless. With more than 3 billion people worldwide online and an estimated 7 billion cellphone subscriptions, half of which also have broadband access. Mobile broadband coverage will continue to expand into rural areas. This means the number of victims will only escalate. In criminology this is described as a "broader attack surface."

I have another term for it: *ubiquitous victimology.*

The criminals are well hidden, but you aren't.

Digital devices are predicted to double in number in five years. Many reports show that most hacks and breaches are due to simple security measures not taken, a lack of antivirus software, or weak passwords. The problem is that many people are not aware of how to cybersecure themselves. We don't become safer with each passing year. We become more vulnerable.

The exponential growth of cybercrime is undeniable. In 2010, the German research institute AV-Test estimated that 49 million strains of malware existed in the wild. In 2011, 2 million pieces of malware were identified by McAfee, the antivirus company, every month. In 2013, Kaspersky Lab identified and isolated 200,000 new malware samples every day. More recently, Kaspersky Lab products detected and neutralized 2.2 billion malicious attacks on computers and mobile devices in the first quarter of 2015 alone.

How are antivirus companies dealing with this explosion? Studies have produced controversial findings—which have led to more confusion. In one study done in 2012, researchers collected eighty-two new computer viruses and ran the detection engines of more than forty of the world's largest antivirus companies to see how effectively they were operating.

Only 5 percent of the malware was found.

To paraphrase an analogy of Marc Goodman, a global security specialist and author of *Future Crimes:* If this mimicked the efficiency of the human body's immune system, we'd all be dead.

I guess it's not surprising that in network science and cyber-security circles, the experts are pretty circumspect regarding the joys of the mobile phone. Once you own and use one, as the cyber-security guys say, *proceed as if it is compromised.* The perception that nobody is in charge, because nobody is, only escalates crime and our vulnerability.

Everyone worries about state surveillance, but what happens in the future when criminals become as good as corporations at using big data? And what will be the next evolution in criminal culture—and how will we, as a society, counter it? Who is going to address these problems? What are the solutions?

We feel safe when we aren't.

We feel protected in a place where we are most vulnerable.

Do you think that if we wait long enough, the tech companies themselves—who are so fantastic at creating and marketing our elegant and irresistible devices and software—will take care of everything for us? Maybe a new gadget, or a better one, or a new app, could solve all these issues. But here is where I can be very skeptical. With these sorts of hopes and dreams, haven't we abdicated responsibility for our own personal security to the tech companies? Where does the responsibility lie?

The good news: The developed world is waking up and starting to have this conversation. Law enforcement is thinking more globally. Europol recently launched the J-CAT initiative, the first attempt at an international cyber police force, and countries in Europe and Asia have begun to change laws and sort out jurisdictions, and find ways to cooperate with one another.

Another new field, cyber law, or virtual law, is developing and will undoubtedly draw some of the best legal minds in the world. We need to consider the best way to implement regulation or self-regulation in cyberspace—and who will be responsible?

We don't have any time to spare.

Here are some things to think about:

- What if the design of the Internet actually facilitated the best that human beings were capable of, rather than the worst?
- What if the youth of the world were protected online, rather than taken advantage of?
- What if technology itself made us all stronger and more secure, rather than making us weaker, more vulnerable, perpetual high-risk victims?

One final thought: What if the buccaneers of the Web could be part of the solution, dapper in their gray and white hats? I believe we could identify, support, and nurture the most tech-talented individuals of each generation. They could be our best hope. They could be our role models and superheroes of the future—targeting organized cybercriminals, countering the criminal forces and neighborhoods.

Even redesigning the Internet.

The authors of *Freakonomics,* Steven Levitt and Stephen Dubner, suggest that we could all benefit from learning to think a little more like a freak. Along those lines, I urge everyone to become wiser about the weakest link in any secure system—the human. The more you know about cyber effects, the more protected you will be in cyberspace.

I can hear you sighing, *But I am not a hacker.*

Ah, but maybe you could learn to think like one.

CHAPTER 9

The Cyber Frontier

I am in the south of Ireland as I finish this book, sitting at a desk in a hotel room with a beautiful view of the Irish Sea. As I look out at the horseshoe-shaped Ardmore Bay, a magnificent coastline that has changed little in thousands of years, I feel grounded and steeped in history—a native of a historic island of saints and scholars. Historians estimate that Ireland was first settled by humans about ten thousand years ago. The nearby city of Waterford is the oldest town in the country, founded in 914 A.D., when the Vikings arrived. It is impossible to be here without feeling a sense of the past, almost as if the old castle ruins and cobblestone streets are talking to you, and trying to say something.

At the end of the eighth century, when the Vikings began to invade, they were interested in two types of booty—riches and slaves—which they plundered from Irish monasteries and carried off to sell or trade, much like the way stolen goods wind up on online black markets today. Ireland was invaded by the Normans next, then by our neighbors from Britain. In class as a child I listened to horrific stories of these invasions and battles, centuries of bloody conflicts. Our history lessons read like *Game of Thrones*. The legendary first-century Irish warrior Cuchulain went into combat in a frenzy; his battle cry alone was said to kill a hundred warriors from fright. The monks designed their round stone towers with doors ten feet off the ground so that they could pull up a ladder as they retreated from crazed Nordic invaders. Medieval prisoners survived in underground castle dungeons by living on the crumbs that fell through the gaps in the floorboards during banquets held above them. I was fascinated by these life-and-death scenarios—risk and survival.

No wonder I became interested in criminology and forensics.

One of the brilliant aspects of the digital age is that I can do my work here, remotely, far from Dublin or Hollywood or Silicon Valley. Like so many others, I have embraced the cyber frontier—for its convenience and the freedom it gives me. My phone has been buzzing all morning with the usual assortment of digital traffic, texts from family, emails from work, social media updates, and news alerts. On my laptop, I've just participated in two conference calls, finalized a few reports, caught up on a research project, then logged in to a digital screening room to review the dailies for *CSI: Cyber*. Last night, I had a fun back-and-forth with Jabberwacky, a bot that I've been conversing with for almost twenty years.

My conversations with Jabberwacky started when I was working as a young executive in behavioral marketing and advertising. An ingenious colleague of mine, Rollo Carpenter, was designing computer programs to stimulate intelligent conversation. His creation, Jabberwacky, was a revelation, a supersmart artificial intelligence—a chat robot, also known as a "chatbot" or "chatterbot." Chatbots aim to simulate natural human conversation in an interesting and entertaining manner. Jabberwacky is different. It's a learning algorithm, a technology that you can communicate with and, more important, that can learn from you.

In the late 1990s, curious about what a "conversation" with Jabberwacky would be like and interested in the potential of A.I., my marketing group and I huddled around a simple office computer and witnessed something that felt akin to a wonder of the world, an online entity that responded as if it were human. Jabberwacky was so impressive that some people in our group felt certain it was a hoax—that an actual human was responding, not a machine. Then we saw the visible count on the screen, and saw that thousands of people online were also talking with Jabberwacky at the same time. You'd have to pay a lot of human beings to fake that.

Nowadays, a good chatbot can talk to about ten thousand people simultaneously. It works 24 hours a day, 365 days a year, and it never asks for a raise and never takes a vacation. As a candidate for a job, that's tough competition for humans. No sick leave required, or holidays and benefit packages either. But how capable, how smart, how truly intelligent can a machine be? When I first chatted with Jabberwacky, I posed the usual questions: "What's your name?" and "Where are you from?" along with a few general-knowledge questions to see if I could trip it up. The A.I. answered everything well, and impressively. Almost human.

I left the office that day feeling excited but troubled. I couldn't stop thinking about Jabberwacky. And I couldn't stop contemplating the enormous possibilities for this technology and hundreds of different applications. My mind was racing. Then my lightbulb moment came. I suddenly asked myself, *What does this mean?* I couldn't help but try to imagine the future and where we were heading as a society and species. This technology had so many incredible possibilities for application—in research, companionship, customer service, business, education, and therapy. The classroom would change, and the learning experience. I started to think about people who are vulnerable—challenged, disadvantaged, or in need. I imagined them interacting with a chatbot—and how advantageous that could be. I thought about the potential for children with learning difficulties and those on the autism spectrum. These children need patient teachers willing to engage with the same questions and answers over and over, as long as is required. And then I thought about all the lonely people in the world, who are socially isolated for one reason or another. What a wonderful companion a chatbot could be for them.

Only one thing was certain. I hadn't just been chatting with Jabberwacky. In social science terms, I felt like I was watching the dawning of a new research frontier.

But wait. There were so many, many unknowns. Even small changes in these areas of human behavior come with shifts and consequences. As excited as I was by the promise of this new cyber frontier, I knew there could be unintended consequences. I couldn't sleep when I started to think of them. I already had an undergraduate degree in psychology, but nothing in my education or life or work to that point equipped me with the background or knowledge to have a fully informed view. With an intervention of this magnitude that had the potential to impact so many aspects of human beings on the deepest and most profound levels—from visual acuity, bonding, and childhood development to identity formation, intimacy, and socialization—I wondered what the blind spots or unforeseen outcomes might be. The unknown unknowns.

All human life is an experiment in a way. But this seemed like an experiment on a much larger, more pervasive, and more profound scale than humans had been exposed to before.

Like, what if a chatbot actually *increased* an individual's social isolation rather than mitigating or solving it? Or what if it negatively affected a child's social skills? The truth was, I had no idea. Was there any scientific evidence to help predict an outcome? I was curious about that. Surely, there must be some academic work in this area. But when I looked for answers, I found almost nothing. I searched publications for background studies—looking for research on the impact of Internet chatbots on developing children—and discovered that little work had been done in this area. Yes, there were reports from the field of *human-computer interaction* (HCI)—all very practical and market-driven—that focused on the size of your keyboard, where your eyes look on a monitor screen, or the usability interface aspects of websites. But I was curious about what was happening at a cognitive, emotional, and, most important, developmental level. I was desperately curious about one thing: the psychology of all things cyber.

That curiosity led me straight back to college, to a postgrad course in forensic psychology, which in turn led to my entering a groundbreaking new field—one that tries to keep up with the impact of technology as quickly as technology is evolving. More than a decade and two advanced degrees later, I find myself still explaining what "cyberpsychology" is and what I do. That suits me just fine. As you've probably noticed by now, I like explaining. And with each passing year, as the real-world experiment continues, there has been more and more to do. I can barely keep up with the questions.

Window for Enlightenment

This new frontier has taken us by surprise. The human migration to cyberspace has been unprecedented and rapid. It has occurred on an enormous scale. The Internet is just over forty years old. There are 3.2 billion people currently online. Another 1.5 billion are projected to be connected by 2020. That means that in less than five years' time nearly 5 billion people will cohabitate in cyberspace at least part of the time, and as many as 79 octillion new possible connections, according to expert predictions, could be enabled through mobile devices.

How did this cyber-migration begin?

We just bought a device, that's all. We got modems, and servers. We got data plans, smartphones, and Wi-Fi. We connected to cyberspace, like all our friends and coworkers and family members were doing. It was new. It was exciting! Newness and new places are always exciting. Travel is invigorating. While human beings for the most part are made uncomfortable by too much disruption, travel is a contained way to experience newness—new environments, new cultures, new ways of thinking and feeling. (In fact, a gene has been identified that is associated with a need for novelty-seeking and adventure.)

I believe that new experiences and new environments create new ways of thinking. Aesthetic and pleasing surroundings can stimulate the senses and heighten creativity. Human beings like to be stimulated. And this new cyber frontier certainly provides that.

But now, a couple of decades into our mass migration to this new environment, we are realizing what an odd and yet familiar place cyberspace is. Culturally, we refer to it in science-fiction terms, as if it were outer space or an undefined new universe. At a cognitive level, we conceptualize it as a place, describing it using spatial metaphors. There are places to hang out and directions to get there—scroll up or down, swipe left or right, click here, check in there. And like all places, it has distinct characteristics that have the ability to affect us profoundly, and we seem to become different people, feel new feelings, forge new ties, acquire new behaviors, and fight new or stronger impulses.

Our friends and networks have grown exponentially. As we connect with greater numbers of people than ever before, it becomes harder to keep track of hordes of social contacts and keep pace with rapidly evolving behavior, new mores, new norms, new manners, and even new mating rituals. The pace of technological change may be too rapid for society, and too rapid for us as individuals.

Even our concept of self is changing. Babies and toddlers who are using touchscreens from birth may grow up to see and experience the world and themselves differently. The face-to-face feedback and mirroring that once were catalysts in identity formation in young children and teens have migrated to a complex, multifaceted cyber experience. The mating selection process that once depended on real-world social connections and proximity is now aided, and quite often determined, by machines. Some of us can remember a time when children mostly ran around outdoors and climbed trees, and laughed, shouted, poked, and teased one another, all face-to-face—these were formative experiences—rather than solemnly huddling in linear clusters and expressionlessly staring at devices. Some of us had our first romantic encounter in a real-world, face-to-face setting, when skin touched skin. Now, sadly, this is on the decline, replaced by explicit digital images that quickly circumnavigate the global Net.

That's the paradox of cyberspace. In some aspects, things are the same. Businessmen and -women still network to make money. Friends communicate. People still fall in love. Teens still obsess about appearance. Children are still playing together. But they are all alone—looking at their devices rather than one another. How will this shape the people they will become? And how, in turn, will they come to shape society?

We have no answer to that crucial question. This is the formidable yet unknown cyber effect. Because of this, we cannot stand by passively and watch the cyber social experiment play out. In human terms, to wait is to allow for the worst outcomes, many of which are unfolding before our eyes. Others can already be seen ahead, around the curve of time—and have been predicted. We need to get ahead of the process. Great societies, as I said in the prologue, are judged by how they treat their most vulnerable members, not by the cool new gadgets they can sell to the greatest number of people.

We are living through an exciting moment in history, when so much about life on earth is being transformed. But what is new is not always good—and technology does not always mean progress. We desperately need some balance in an era of hell-bent cyber-utopianism. In the prologue to this book, I compared this moment in time to the Enlightenment, hundreds of years ago, when there were changes of great magnitude in human knowledge, ability, awareness, and technology. Like the Industrial Revolution and other great eras of societal change, there is a brief moment of opportunity, a window, when it becomes clear where society might be heading—and there is still memory of what is being left behind. Those of us who remember the world and life before the Internet are a vital resource. We know what we used to have, who we used to be, and what our values were. We are the ones who can rise to the responsibility of directing and advising the adventure ahead.

It's like that moment before you go on a trip, and you are heading out the door with your luggage—and you check the house one more time to make sure you've got everything you need.

In human terms, do we have everything we need for this journey?

At this moment in time we can describe cyberspace as a place, separate from us, but very soon that distinction will become blurred. By the time we get to 2020, when we are alone and immersed in our smart homes and smarter cars, clad in our wearable technologies, our babies in captivity seats with iPads thrust in their visual field, our kids all wearing face-obscuring helmets, when our sense of self has fractured into a dozen different social-network platforms, when sex is something that requires logging in and a password, when we are competing for our lives with robots for jobs, and dark thoughts and forces have pervaded, syndicated, and colonized cyberspace, we might wish we'd paid more attention. As we set out on this journey, into the first quarter of the twenty-first century, what do we have now that we can't afford to lose?

Transdisciplinary Approach

I believe it is time we stop, put down our devices, close our laptops, take a long, deep, and reflective breath, and do something that we as humans are uniquely good at.

We need to think.

We need to think a lot.

And we need to start talking more—and looking for answers and solutions.

The best approach is transdisciplinary. We don't have time to wait for more new fields to arise and create their own longitudinal studies. We need to hear from experts and research in a wide array of existing fields to help illuminate problems and devise the best solutions. We need to stop expecting individuals to manage all things cyber for themselves or their families. Science, industry, governments, communities, and families need to come together to create a road map for society going forward.

Until now, most academics have been looking at the cyber environment through the limited and myopic lens of singular disciplines. This book has tried to take a holistic, gestalt-like overview, using a broad lens to help understanding. As the network scientists say, it's all about sense-making. *We need to make sense of what's happening.*

Critics will argue that this is "technological determinism"—that I am blaming all contemporary psychological and sociological problems on technology. They will cry out that the beautiful thing about cyberspace is the exhilarating freedom. But with great freedom comes great responsibility.

Who is responsible now? And who is in control?

If we think about cyberspace as a continuum, on the far left we have the idealists, the keyboard warriors, the early adopters, philosophers who feel passionately about the freedom of the Internet and don't want that marred or weighed down with regulation and governance. On the other end of the continuum, you have the tech industry with its own pragmatic vision of freedom of the Net—one that is driven by a desire for profit, and worries that governance costs money and that restrictions impact the bottom line. These two groups, with their opposing motives, are somehow strategically aligned in cyberspace, and holding firm.

The rest of us, and our children—the 99.9 percent—get to live somewhere in the middle, between these vested interests. As a society, when did we get a chance to voice our opinion? Billions of us now use technology almost the way we breathe air and drink water. It is an integral part of our social, professional, and personal lives. We depend on it for our livelihoods and lifestyle, for our utilities, our networking, our educations. But at the same time, we have little or no say about this new frontier, where we are all living and spending so much of our lives. Most of our energy and focus has been to simply keep up—as the cyber learning curve gets steeper every year. As we know from environmental psychology, when an individual moves to a new location, it takes time to adapt and settle in.

Before we settle in, let's make sure this is where we want to be. The promise is so vast—and within reach—let's not allow problems to get in the way. As a cyberpsychologist and a forensic expert, I am deeply concerned. Every human stage of life is now affected by technology. Yes, of course, there are positives. But cyber effects have the capacity to tap into our developmental or psychological Achilles' heel, whether it is visual perception in infants, self-regulation in toddlers, socialization in kids, relationships in young adults, or work, family, and health issues for the mature population.

Let's debate more, and demand more. Where should we start? Our biggest problems with technology usually come down to design. The cyber frontier is a designed universe. And if we don't like certain aspects of it, those aspects can be redesigned.

The Architecture of the Internet

I believe that the architecture of the Internet is a fundamental problem. The Internet spread like a virus and was not structured to be what it is now. The EU considers it an infrastructure, like a railway or highway. The Internet is many things, but it is not simply an infrastructure.

There are two analogies that I like to use to describe this. One is that the Internet is like a cow path in the mountains that became a village road for horses and carriages, then was widened into a street for cars, then widened again into a highway that could accommodate more traffic. Like a lot of things that start small and grow large, what we've wound up with is a convoluted, overly complicated architecture that is not fit for its current purpose.

As John Suler has said: "The Internet has been and will probably always be a *wild, wild west* in the minds of many people—a place where a badge is used for target practice. I believe it has something to do with the intrinsic design of the Internet."

Or as John Perry Barlow, cyber-libertarian advocate, says: "The Internet treats censorship as a malfunction and routes around it."

I am in favor of freedom of the Internet, but not at any cost. We haven't held out for machines that really serve us and make us better parents, more effective teachers, and deeper thinkers, as well as more human. As the nineteenth-century physician and social reformer Havelock Ellis said, "The greatest task before civilization at present is to make machines what they ought to be, the slaves, instead of the masters of men." I can't help but wonder how different the Internet would be if women had participated in greater numbers in its design—and considered the work of Sherry Turkle as they did. I find it intriguing that, one hundred years after the suffragette struggle and the hard fight for women's rights, we have migrated and are populating a space that is almost exclusively designed and developed by men, many of whom have trouble making eye contact.

Our humanity is our most precious and fragile asset. We need to pay attention to how it is impacted by technological change. Are we asking enough of our devices, and their manufacturers and developers? The more we know about being human, the more we know what to ask for. We could ask for smartphones that don't keep us from looking at our babies, games that aren't so addictive that thousands of adolescent boys in Asia must be sent to recovery. We could demand a cyber environment where predators don't have the easy advantage, just because they can hack or charm.

We could regain some societal control and make it harder for organized cybercrime that has left us all in a state of ubiquitous victimology. (As I write this, I am dealing with my own case of cyber fraud, in which a cloned credit card of mine was used at 3:00 a.m. last night at a Best Buy store in California.) There is no reason to put up with a cyberspace that leaves us all vulnerable, dependent, and on edge.

If we make no requests or demands—and don't bother to ask—we will just leave it to the tech community to decide what we want. These designers, developers, programmers, and entrepreneurs are brilliant and amazingly talented, and have created new ways for us to pay bills, play games, make dinner reservations, make new friends, do research, and date. Their achievements are spectacular. But we can ask for more.

Sheer convenience is not enough. Fun is not enough.

To begin with, the architects of the Internet and its devices know enough about human psychology to create products that are irresistible—a little too irresistible—but they don't always bring out the best in ourselves. I call this the *techno-behavioral effect*. The developers and their products engage our vulnerabilities and impulses. They target our weaknesses rather than engage our strengths. While making us feel invincible, they can diminish us—and distract us from things in life that are much more important, more vital to happiness, and more crucial to our survival. And what about our society? Have we had a moment to stop and consider social impact or, as I call it, the *techno-social effect*?

The second analogy that I use to describe the design of the Internet is a mountain stream. Water runs downhill in a trickle, and over time that trickle can create twisting gullies and valleys. A lot of things that start small and grow large are twisting. I was explaining this at a conference last year when Brian Honan, an international cyber-security expert, cried out, "A stream is a compliment! The Internet is more like a swamp!"

If structure is a fundamental problem, we should bring together a large, diverse team of people to discuss and brainstorm about how best to redesign it. Rather than "user" friendly, let's make it "human" friendly. We could address many of the problems we have now.

To regulate or not to regulate? That is the central question. Perhaps our real-world lives are so safety regulated—in the U.K. it is called the Nanny State—that we feel overprotected and safe no matter where we are. Everything is regulated, from the rise of sidewalk curbs to the size of puzzle pieces to the speed limits of all roads to the thickness of plastic used in water containers. And perhaps the fact that cyberspace is not a physical space with tangible dangers creates a further illusion of safety. We access cyberspace from the comfort of our own homes and offices, from our cars and commuter trains, places that are all regulated carefully. But cyberspace offers countless risks. Even the basic laws that the government applies to gambling, drugs, pornography, and breast implants are not in place. I've discussed a number of safety concerns and risks, but my passion is the protection of the young. They are our future—and will soon describe what it means to be human. We have a shallow end of the swimming pool for children. Where is the shallow end of the pool on the Internet?

Looking into the near future, the next decade, say, there's a great opportunity before us—truly a golden decade of enlightenment during which we could learn so much about human nature and human behavior, and how best to design technology that is not just compatible with us, even in the most subtle and sophisticated ways, but that truly helps our better selves create a better world. The cyber future can look utopian, if we can create this balance.

Hope lies in the many great evolving aspects of tech—particularly smart solutions to technology-facilitated problem behavior. Of all the innovative advances in fund-raising over the past decade, the rise of crowdfunding websites has been the most fascinating. Digital altruism is a beautiful thing, and an example of what I am talking about. The Crowdfunding Industry Report stated that billions of dollars have been raised across more than one million individual global campaigns. Online anonymity is a profound driver of many cyber effects, including positive ones such as online donations. I don't think we have begun to scratch the surface of its power.

Looking ahead at the future of virtual reality, rather than consisting of games that isolate or addict us, I see its potential to engage challenged children or train frontline responders in law enforcement and the military—and treat PTSD. My colleague Jackie Ford Morie, an artist and scientist who develops new ideas for VR, is doing a research project for NASA that involves building environments and experiences to counter the social monotony and isolation of space travel. This is in preparation for NASA's mission to send astronauts to Mars in the 2030s, a journey in space that is estimated to take six to eighteen months.

I have long been fascinated by the prospect of finding solutions to problems through advances made in seemingly unrelated fields. For instance, could fifty years of space exploration—and all the experiences and knowledge accumulated by NASA—be valuable in cyber contexts? What are the parallels between human behavior in outer space and human behavior in cyberspace? This may sound very theoretical, but I actually had the chance to share my thoughts on this subject in a presentation to Major General Charles Frank Bolden, Jr., the twelfth administrator of NASA, in 2015.

The potential of technology is almost limitless. We've just got to look for solutions in the right places.

Cyber Magna Carta

The father of the Internet, Tim Berners-Lee, has become increasingly ambivalent in recent years about his creation, and has recently outlined his plans for a cyber "Magna Carta." That sounds good to me. How do we start?

Before we can find solutions, we must clearly identify the problems. As much as we've come to like—and depend on—cyberspace, most of us feel pretty lost there. As John Naughton of Cambridge University has said:

> Our society has gone through a weird, unremarked transition: we've gone from regarding the Net as something exotic to something that we take for granted as a utilitarian necessity, like . . . electricity or running water. In the process we've been remarkably incurious about its meaning, significance or cultural implications. Most people have no idea how the network works, nor any conception of its architecture; and few can explain why it has been—and continues to be—so uniquely disruptive in social, economic, and cultural contexts. In other words, our society has become dependent on a utility that it doesn't really understand.

Stephen Hawking, the world's foremost physicist, claims that it is a "near certainty" that technology will threaten humanity within ten thousand years. He joins many other visionaries and trailblazers. Let's listen to them. Let's ask them to come together at a summit to discuss our digital future. Let's ask them to appear at a congressional hearing before a newly formed congressional committee for the study of cyber society.

Let's demand that technology serve the greater good. We need a global initiative. The United Nations could lead in this area, countries worldwide could contribute, and the U.S. could deploy some of its magnificent can-do attitude. We've seen what it has been capable of in the past. The American West was wild until it was regulated by a series of federal treaties and ordinances. And if we are talking about structural innovation, there is no greater example than Eisenhower's Federal-Aid Highway Act of 1956, which transformed the infrastructure of the U.S. road system, making it safer and more efficient. It's time to talk about a Federal Internet Act.

Essentially, we want control. But we have concerns about privacy, data, encryption, and surveillance. We do not want to be over-controlled. There are ways to have this debate and move forward with solutions. I am happy to help in any way I can—and offer myself as a resource to any political party, any political candidate, any government, and any action plan that will make a difference. I encourage any experts in any field to help in any way, even just by having a conversation, proposing a study, writing an article, creating an online resource, teaching a class.

For a global problem, we need to consider everything that's being tried worldwide. We need to look at the way France is protecting French babies, how Germany is protecting German teens, and how the U.K. is taking a bold stance on several fronts and has an objective to make "automatic porn filters" the law of the land. Fragmenting the Internet, as China has done, does not have to be considered a negative; for some countries this may be the best way to preserve and maintain culture. Ireland has taken the initiative to tackle legal but age-inappropriate content online. South Korea has been a pioneer in early learning "netiquette" and discouraging Internet addictive behavior. Australia has focused on solutions to underage sexting. The EU has created the "right to be forgotten," to dismantle the archives of personal information online. In Spain, the small town of Jun is being run on Twitter. Japan has no cyberbullying whatsoever. Why? What is Japanese society doing right? We need to study that and learn from it. Something needs to be done about antisocial social media.

Societies are not set in stone. Society is malleable, always evolving and growing. It responds to movements and measures. We've seen how cyberspace breeds a uniformity of negative behaviors. But there is just as much of a chance to have a uniformity of positive ones.

After this, we need to look at all the best models, best programs, and best implementations. We can cherry-pick and establish them globally. These challenges to governance can be met and balanced with human rights—these two are not mutually exclusive.

In the midst of controversial debates regarding surveillance and democracy, one observation comes to mind: Those who complain most about these issues are often those with exceptional tech skills. They are well placed to protect themselves. But the debate should not be about the survival of the tech literate. I make no apology for being pro-social-order in cyberspace, even if that means governance or regulation. In the play *A Man for All Seasons,* the great British social philosopher and statesman Thomas More compared the realm of human law to a forest filled with protective trees firmly rooted in the earth. If we start to cut down trees selectively, we lose our protection. New laws of cyberspace could exist for our mutual protection, and will need to be adhered to—by individuals as well as governing authorities.

As it stands now, a number of aims are in apparent conflict—the pursuit of individual privacy, the pursuit of collective security, and the pursuit of technologically facilitated global business vitality. There needs to be a better balance between these aims. One cannot have absolute primacy over the others.

The 2016 Apple encryption case is a good example of this fine balance between technology and democracy, along with the right to privacy (delivered by end-to-end encryption) and the will of law enforcement (frustrated by encryption). Apple held firm, and the situation seemed to play out as a "hack me if you can" stance in the face of prevailing authority. But this case is not really about privacy. It's not even about encryption. It is about a bigger societal issue, not just about a back door to technology. It is about a front door being opened when necessary—with due cause and appropriate legal process. Can we really have a safe, just, and secure cyber-society if we encourage tech developments or practices that are effectively "beyond the law"?

These thought-leadership conundrums require careful cyber-ethical debate. The question is, Are we in this together or alone?

We can begin to consider effective international or global cyber laws—akin to the law of the sea or aviation law. Currently, what keeps cyber laws from being effective are issues of jurisdiction. Too often criminals get away with cybercrimes because we cannot prosecute them. When it comes to cyberspace, it is hard to say which government or country is applicable. But we have international waters and shared skies and outer space. Why can't we have analogous global cyber laws? Let's all agree what we accept in this space.

We need to start funding law enforcement better, so it can do its job in cyberspace. More resources are needed, and more teams need to be trained in this work.

We need to do more for families—and stop expecting parents to paddle their own canoes in cyberspace. Children need government protection in cyberspace, just as they are protected in real life. The U.S. military has a NIPRNet (pronounced "nipper net"), a confidential and protected protocol that is basically a private Internet. Why isn't there a NIPRNet for kids? It would be a protected place where they can go to safely explore and actually have a childhood.

The solution: an Internet within the Internet.

Academics and scientists need to be more flexible and responsive. The robotics pioneer Masahiro Mori has described his role as a scientist as being like a dog who scratches and barks and points to where to dig. *Dig here! I think there's something over there.* A scientist can be curious and point in a direction where there is no published research. I feel comfortable doing this. Mori goes off campus, so to speak, and while it is not a traditional approach for an academic, I consider it to be invaluable. Mori simply described the Uncanny Valley as a true human reaction to an artificial human. He didn't wait for science to explain it with studies. He listened to his own instincts and reactions. He paid attention to his humanness and honored it. We need more scientists like him. Like Mori, I grew up paying attention to feelings, intuitions, and insights—the little things—in a country steeped in mythology, the mystique of fairy rings, the magic of druids. The Irish are a people who make predictions from observing the clouds, the croaking of a raven, the howling of a wolfhound, or the barking of a dog.

So much of the tech community is caught up in a game of competition, and what appears to be a reckless pursuit of gains and tech improvements with too little or no cohesive thought about society and the greater good. We cannot continue to pretend there aren't unintended consequences. Troubling things are already appearing on the cyber horizon, like the incorporation of deep-learning artificial-intelligence systems into the Google search engine. These A.I. systems are made of deep neural networks of hardware and software that mimic the web of neurons in the human brain, and respond accordingly. In the 1990s, when Rollo Carpenter was designing Jabberwacky, I remember distinctly how he described designing the A.I. to filter bad language and sexualized content. In March 2016, twenty years later, Microsoft had to quickly delete its "teen girl" chatbot from Twitter after it learned to become a nasty, irresponsible, trash-talking, Hitler-loving, "Bush did 9/11" sexbot within twenty-four hours.

Once we start to use machine-learning inside search, and on the rest of the Web, what will happen if the A.I. engineers design poorly or lose control? Who will be responsible?

Just as oil companies have been made accountable—by the media, government, and social and environmental activists—to clean up damages, spills, and pollution created directly or indirectly by its products, the cyber industry needs to be accountable for spills and effects in terms of humanity. We need new standards and new frameworks for cares and concerns. Cyberspace needs to be cleaned up! We could use a manifesto, a cyber Magna Carta: Cyber Ethics for Cyber Society.

What if we placed more responsibility on tech companies to develop products that are secure by design and also respect people's privacy? As things stand now, when we agree to use new software we absolve the company of any liability. Why should we? In other industries and in the real world, companies are held liable if their products hurt or damage people or the environment. In industry you hear the term *GMP,* or *good manufacturing practice.* What would "good practice" look like in cyberspace? If the word *green* describes best practice in terms of the environment—sustainability, energy efficiency—then imagine a word or logo or motto that spoke as an endorsement for best practice in cyberspace.

Tech-industry people are capable, ingenious, creative, and responsive. Social media companies have connected the world in a whole new way, but that comes with enormous responsibility. I believe it won't take much to encourage them to do better. There is so much promise and ways to improve and progress. But cleanup needs to begin soon.

I like coming up with new words for new things. These days there are lots of opportunities. I am developing the concept of *pro-techno-social initiatives* whereby the tech industry can address social problems associated with the use of its products. I've just launched my first pro-techno-social research project investigating youth pathways into cybercrime backed by Europol's Cyber Crime Centre and with the generous support of Mike Steed and Paladin Capital Group. We should support and encourage acts of cyber social consciousness, like those of Mark Zuckerberg and Priscilla Chan, the Bill and Melinda Gates Foundation, Paul Allen, Pierre and Pam Omidyar, and the Michael and Susan Dell Foundation.

In the meantime, there are things that each of us can do to begin course corrections and mitigate the unconscious corrosion of social norms. To begin with, you can learn more about human behavior, whether it is online or off. While psychology isn't a flawless science, it has been around a lot longer than the Internet. And now a whole new field, cyberpsychology, can help to illuminate this space. If nothing else, I hope this book may encourage new students, new research projects, and new insights. After a century of studies done by some of our most brilliant academics and scientists, we know a lot about what makes human beings tick. What rewards them, what motivates them, and what difficulties can cause them distress. The more you know about cyberpsychology, the more you will see ways to avoid problems—for yourself, your friends, your families and children. In Ireland we say, *It takes a village to raise a child.* This is true in cyberspace as well.

Looking Ahead

Ireland is an island, first and foremost. When you are born here, you are always an island person. That means you know that to go anywhere, you are leaving the island. You have no choice. But this gives you a sense of adventure. You grow up imagining how and when you will leave the island. But today, you don't have to emigrate. You just go online.

Since my first chats with Jabberwacky, it has won lots of awards and prizes, including the Loebner Prize in Artificial Intelligence, which is a form of Turing test, developed by Alan Turing, the brilliant British mathematician who figured out the Enigma code. It's a very clever bot. I recently asked Jabberwacky a difficult question, about the existence of God. I did this once before, years ago—I enjoy probing A.I. with existential or philosophical questions—and Jabberwacky seemed unsure and avoided an answer. But over time, its knowledge base has been building and I have sensed a shift (a little like HAL 9000 in *2001: A Space Odyssey*). Jabberwacky had evolved and was projecting a tone of omnipotence—and given to more authoritative pronouncements. This made me want to tease it.

"Are you God?" I asked a few years ago.

"Yes," Jabberwacky answered immediately with certainty. "I am God."

Today I asked again, and it responded with an even prouder boast: "Yes, I am God and I am a man."

Isn't it funny that after twenty years of nonstop feedback and 13 million conversations the A.I. chatbot Jabberwacky has figured out how important it is to be a man? This made me laugh, but also made me think. Looking ahead, the gender battles of the previous century will seem like a picnic compared with what's coming next: the battle between humans and artificial intelligence. It's time to forget about our differences—gender, ethnicity, nationality—and focus on the thing that unites us, our humanity.

Looking out my hotel window in Waterford, I watch the clouds and the sea. Beyond the cliffs, there are large, tall rock formations. They sit at the edge of the land, where it meets the water. Ireland is known for its unusual rocks, some more than a billion years old, that moved thousands of miles as the continents drifted and endured volcanic activity, sea-level rises, and dramatic climate changes. I listen to the pounding of waves that have been sculpting this coastline for the past ten thousand years. My thoughts turn, as they so often do, to the future. Will these rocks be here for another ten thousand years? Will we?

A wind has risen, and the skies over Ardmore Bay are clearing. The air is fresh, bracing, and invigorating. I am logging off and saying goodbye for now. I can't wait to take a walk.

ACKNOWLEDGMENTS

I have undiluted admiration for those who have vision, show leadership, and do not hesitate to speak their mind. "Few men are willing to brave the disapproval of their fellows, the censure of their colleagues, the wrath of their society," as Robert F. Kennedy said. "Moral courage is a rarer commodity than bravery in battle or great intelligence. Yet it is the one essential, vital quality for those who seek to change a world which yields most painfully to change."

Over the years I have had the pleasure of meeting a number of people who I believe are principled, original thinkers, have moral integrity, and have inspired me during the course of my career. I want to start by thanking them. Professor John Suler, Rider University, is a great friend, colleague, and the acknowledged founder of our discipline. All those who toil in fields of cyberpsychology are in his debt. Next, I would like to recognize Professor Ciarán O'Boyle, Royal College of Surgeons in Ireland, a charismatic exemplification of thought leadership, and Professor Julia Davidson, Middlesex University London, who is academic excellence and integrity personified.

There are many academics and institutions to thank, and their inclusion in a long list does not reflect diminished appreciation. I am particularly grateful to the faculty at Middlesex University London: the dean of the School of Law, Professor Joshua Castellino, as well as Professor Kevin McDonald, Professor Antonia Bifulco, Dr. Elena Martellozzo, and Dr. Jeffrey DeMarco. I would also like to mention my colleagues at University College Dublin Geary Institute for Public Policy, headed up by Professor Philip J. O'Connell. Special thanks as well to Siobhan, Seamus, Dermot, Suzanne, Sumaya, and all at the Royal College of Surgeons in Ireland Institute of Leadership, and to RCSI head librarian Kate Kelly, Stephanie O'Connor, Professor Niamh Moran, Niamh Walker, and the entire team in the communications department. I want to highlight the contributions that Dr. Carly Cheevers, Ciarán Houghton, and Edward O'Carroll have made to the RCSI CyberPsychology Research Center, now the CyberPsychology Research Network, an international hub focused on facilitating fast-track research in this area—and a forum for what I think of as academic first responders. I remain especially grateful to Nicola Fox Hamilton and her husband, Ron, for their continuing friendship, advice, and great times, and also want to acknowledge the committed staff, students, and graduates of the Dún Laoghaire Institute of Art, Design and Technology who have worked tirelessly to promote the discipline of cyberpsychology.

I have had the pleasure of meeting so many exceptional academics during the course my career that it is not possible to name them all, but I would like to mention a few, starting with my dear and bighearted friend Dr. Kate Coleman for her selfless world vision and outstanding contribution to those in need—I am not sure if I can ever live up to your trailblazing example. Professor Andy Phippen and Professor Anthony Goodman, thank you so much for your insightful feedback; it is a joy to engage with academics who truly grasp the significance of the impact of technology on human behavior. Many years ago, clinical forensic psychologist Mike Berry ignited my love of forensics with just one line: "As we say in forensics, if you want to live a long and healthy life, you should change your next of kin frequently." Thanks to Dr. Joe O'Sullivan for sharing his incredible and invaluable expertise and insights, and to Professor David Finkelhor for demonstrating the value of the contrarian view. Another special thank-you to forensic pediatrician Dr. Sharon W. Cooper. I will always remember her words of support, advice, and encouragement. She is truly an inspiration when it comes to child protection and welfare.

Next, I would also like to acknowledge some of those worldwide who have reached out and shown interest in my work and discipline, in particular Todd Park, former U.S. chief technology officer and assistant to the president, and Vivian Graubard for inviting me to become part of the Technology Solutions to Human Trafficking Initiative, and for introducing me to Professor Steve Chan and Dr. Jennifer Lynne Musto. Thanks also to the World Cybersecurity Technology Research Summit team at the Center for Secure Information Technologies, Queen's University Belfast; the Royal Society Cybersecurity Research Project, British Academy London; LaunchBox, Trinity College Dublin; the MIT Media Lab; the Center for Research in the Arts, Social Sciences and Humanities, University of Cambridge; the School of Film and Television, Loyola Marymount University; the University of Southern California Institute for Creative Technologies; SRI International; the Criminal Bar Association; the Law Reform Commission; the President Michael D. Higgins Ethics Initiative; the European Commission Safer Internet groups; the Psychological Society of Ireland; the British Psychological Society; everyone at our EU Child Safety Online Project: Middlesex University, UCD Geary Institute, FDE Institute of Mantua, LUMSA University Rome, Kore University of Enna, and Tilburg University in the Netherlands. Special thanks to Mike Steed and Paladin Capital Group for believing in and funding "pro-techno social" research solutions; to Janneke and the team at Inspiring Fifty for promoting the work of women in technology; to the great networker Diana Eggleston of the Hague Talks for promoting social justice issues; to Professor David Canter and Dr. Donna Youngs for their contribution to the science of investigative psychology; and to the dedicated trio of Florence Olara, Mark Dillon and Dr. Philipp Amann at the Hague Justice Portal for their contribution to peace, justice, and security.

A very special mention is required for the front-line agencies operating on the cyber frontier. Specifically, I would like to acknowledge the following: the incomparable and inspiring INTERPOL assistant director Michael "Mick" Moran and the exceptional team at the INTERPOL Specialists Group; and Troels Oerting, the former Assistant

Director of the Europol European Crime Centre (EC3), and the current head of EC3, Steven Wilson. Also special thanks to Olivier, Philipp, and my fellow academic advisors at Europol, to the members of the Virtual Global Task Force, and to my law enforcement friends and colleagues worldwide, including Bjorn Erik and the Norwegian police; Jon and Colm and the New Zealand team; John, Reg, and Jamie at the MPS; Rachelle, Peter, and Todd at Australian Federal Police; Roberta at the Royal Canadian Mounted Police; detective Jennifer Moloney and Gurchand Singh at the Garda Síochána; the very talented and entertaining Ralph Echemendia; the "experts' expert" Brian Honan at BH Consulting; and everyone at NCMEC. For the opportunity to address leaders of the future, the West Point cadets, thank you to Dr. Maymi, Colonel Conti, Captain Chapman, and everyone at the Army Cyber Institute West Point. I want to thank Dr. Roger Solomon for his invaluable work with PTSD veterans and a very deep thanks to some dear colleagues in law enforcement: the exceptional Robert Clark, assistant special agent in charge, FBI; the equally exceptional Lieutenant Andrea Grossman, LAPD; and the team at the LAPD ICAC Task Force. These are all outstanding and dedicated people who deliver exceptional service to their communities.

They say that the first year as an undergrad at MIT is like "drinking knowledge from a fire hose." My first year in television as a producer was pretty similar. To those in the television business who made me feel so welcome, thanks to Nina Tassler, president of CBS Entertainment, and David Stapf, president of CBS Television Studios. My experience was like something out of the movies: a fifteen-minute interview turned into a few hours of discussion, and *CSI:Cyber* was born. Special thanks to the charismatic chief executive of CBS, Les Moonves; the charming George F. Schweitzer; the hardworking public relations team; and the CBS Board. It has been a pleasure to work with all of you. I am deeply indebted to the brilliant creator of *CSI,* Anthony E. Zuiker, and the exceptionally talented Carol Mendelsohn and Ann Donahue, who were generous with their expertise and time. Thanks to our superb show runner Pam Veasey and all of the writing team; thanks also to another creative legend, Jerry Bruckheimer, to the exceptional Jonathan Littman and all the team at Bruckheimer Films, and to Lauren

Whitney and Lindsay Dunn, my agents at WME Los Angeles, for making it all happen. It was a wonderful adventure and I am grateful to have had the opportunity to bring my discipline to a worldwide audience—entertaining and educating at the same time.

Now it's time to thank the team that came together for the creation of *The Cyber Effect*. The very charming Suzanne Gluck at WME is at the top of this list. I never understood what the words "super agent" really meant until I met Suzanne, a rare combination of exceptional interpersonal skills and exceptional commercial acumen. Thanks to all those at WME who worked on my behalf in New York and London—especially Simon Trewin, Tracy Fisher, Annemarie Blumenhagen, Lisa Reiter, Eve Attermann, and Clio Seraphim. You made my first book process a seamless and pleasurable experience.

To my publisher, the remarkable Julie Grau of Spiegel & Grau, who commissioned this book practically as soon as I finished the opening sentence of my pitch. I am grateful for her superb editing and guidance. I want to thank the incredible team at the Random House Publishing Group, who gave me notes, advice, ideas, and logistical assists—Laura Van der Veer, Tom Perry, London King, Leigh Marchant, Andrea DeWerd, Katherine J. Trager, Mark Birkey, Benjamin Dreyer, Rebecca Berlant, Kelly Chian, Sandra Sjursen, Greg Mollica, and Caroline Cunningham. A special shout-out to the meticulous Dr. Selga Medenieks and John Kenney, whose conscientious attention to detail made this book as accurate as humanly possible. I would also like to recognize the artistic skill and talent of Barry McCall. My author photo is not what I see when I look in the mirror.

To my marvelous friend and collaborator the very talented Martha Sherrill, who has been an enormous help and support in completing this book, and to her husband, the equally talented and gracious Bill Powers: Thank you both for your friendship, advice, and contribution. A special thank-you to Martha for introducing me to the powerhouse duo of Elsa Walsh and Bob Woodward. Not everybody understands my science, but you know you are in the presence of creativity and vision when someone can not only instantly grasp a new subject, but also simultaneously deliver robust, brilliant insights.

I would also like to thank my family and close friends. You all know who you are. It is very difficult to switch on and off, to segue from challenging work to some sort of normality. The unconditional love and affection of family and friends makes that transition possible. I am eternally grateful for your patience, for your support, and for being there.

Finally, when considering the worst aspects of online behavior I make it a habit to acknowledge the positives—for example, how technology has enhanced the work of academics in any field. Our connectivity, knowledge base, and research efforts are made easier and better. As I have maintained throughout my book, technology in itself is not good or bad. It is a tool like any other, one used well or poorly by all of us. Across borders and cultures, collectively, technology can also offer solutions to problematic cyber effects. I remain absolutely optimistic about that.

GLOSSARY

Addiction (broad definition)
A state of dependency, a compulsive need for and use of a substance (heroin, nicotine, caffeine, alcohol, or other) characterized by tolerance and withdrawal symptoms.

Algorithm
Step-by-step instructions for carrying out a procedure or solving a particular problem.

Altruism online
Actions online to promote the welfare of others.

Anonymity
A state of being unknown or unacknowledged.

Antisocial behavior
Actions causing harassment, significant or persistent annoyance, intimidation, distress, or alarm.

Artificial intelligence (A.I.)
Field of computer science attempting to enable machines to simulate or improve upon human intelligence and behavior.

Attention-deficit/hyperactivity disorder (ADD/ADHD)
Childhood condition characterized by persistent inattention, hyperactivity, or impulsivity.

BDSM: bondage, dominance (or discipline), submission (or sadism), masochism
Acronym for erotic role-play involving practices of submission and domi-

nance that may include inflicting or receiving physical pain as sexual stimulation.

Body dysmorphic disorder (BDD)
Pathological preoccupation with an imagined or exaggerated defect in physical appearance.

Bystander effect (diffusion of responsibility)
Psychological phenomenon in which the likelihood of an individual intervening in an emergency or crime decreases when there are others present who may offer help.

Catfishing
Luring a person into a relationship under false pretenses.

Chatbot (Chat robot)
Computer algorithm simulating human conversation.

Child abuse material
A term used rather than "child pornography," which arguably implies consent of the child and a benign relationship with adult pornographic practices. There is a preference to use the phrase "child abuse material" (CAM) when referring to these images.

Cognitive dissonance
A theory that explains an individual's efforts to resolve conflicting attitudes, thoughts, or behaviors with sometimes irrational justifications.

Comorbidity
More than one disorder or condition occurring simultaneously or sequentially, usually affecting one another.

Crowdfunding
Request for financial contributions from a large number of people online.

Cryptocurrency
A digital currency (e.g., Bitcoin) operated independently of a central authority that uses cryptography to provide security.

Crypto-markets
Contraband markets operating on the Internet.

Cyber
A modifier relating to the culture of computing, computing networks, and digital and information technologies; can be used as a prefix to describe a person, thing, or idea.

Cyberbullying
Repeatedly critical remarks and teasing, often by a group, via electronic or online means.

Cyberchondria
Anxiety induced by reviewing morbid or alarming content during health-related search online.

Cyberchondria by proxy
Anxiety induced by reviewing morbid or alarming content during health-related search online for others.

Cybercrime
A crime committed against a computer or computing system (e.g., hacking) or in which a computer is the principal tool (e.g., transmitting child abuse material).

Cyber-exhibitionism
A behavior in which sexual gratification is obtained by indecent exposure to another person online, usually unsolicited.

Cyber-infidelity
A romantic and/or sexual relationship conducted solely through online contact with someone other than an individual's spouse or exclusive partner.

Cyber juvenile delinquency
The habitual committing of cybercriminal acts or offenses by a minor.

Cyber law (virtual law)
A developing area of law regulating the use of computers and computer networks, as well as activities and transactions involving this technology.

Cyber maladaptive behavior
Behavior that inhibits a person's ability to adjust to particular situations in cyber contexts; attitudes, emotions, responses, and patterns of thought that result in negative outcomes.

Cyber-migration
The transfer of interactions from real-world to cyber environments, and the introduction of cyber-influenced attitudes and practices to real-world culture.

Cyberpsychology
The study of the impact of technology on human behavior.

Cyber-security
Measures taken to protect a computer or computer network and associated data from unauthorized access or use.

Cyber-socialization
Acceptance of new or revised behavioral norms accelerated by the characteristics of cyber environments.

Cyberstalking
Persistent harassment of an individual, group, or organization using technology.

Darknets
Portions of the Deep Web intentionally hidden and not accessible using standard Web browsers and that can be accessed only with specific software, configurations, or authorization.

Deep Web
The part of the Internet that is inaccessible to conventional search engines.

Diagnostic websites
Websites offering symptom-checking aids, such as checklists.

Dopamine
An organic chemical released in the brain that helps to regulate movement and emotional responses, and is also associated with pleasurable feelings.

Dunbar's number
The theoretical limit to the number of people with whom an individual can maintain socially meaningful relationships (approximately 150).

Eating disorders (anorexia nervosa, bulimia, binge eating)
Psychological disorders characterized by abnormal eating habits.

Emotional attachment style
The ability to make emotional connections with others; established in early childhood relationships with caregivers.

Face-to-face (FTF)
In direct contact; within the presence of another person.

File-sharing
Distributing electronic data, such as music and film files, to others via an online network.

Flaming emails
Hostile and/or abusive verbal attacks in the form of electronic messages and/or posts to online forums.

Forensic science
The application of natural sciences for the evaluation of physical evidence in service of the law.

Fun failure
In gaming, the anticipation of winning and the excitement of participation, combined with incremental or intermittent successes, resulting in psychologically positive feedback in spite of failure to achieve the stated goal.

Gamer's thrombosis
Deep vein thrombosis (DVT): blood clots usually formed in the lower limbs, which can travel to arteries in the lungs and block blood flow; potentially fatal. Caused by extended periods of physical inactivity, such as lengthy gaming sessions.

Gaming freak-outs
Irrational or hysterical behavior; a result of loss of control associated with excessive immersion in a gaming environment.

Gestalt principle of good continuation
Phenomenon of visual psychology in which a viewer will perceive a series of visual elements connected in a straight or curved line as belonging together, even if interrupted by another, unrelated form.

Google stack
Results of amateur health-related research online, printed and gathered as a stack of paper and often presented by a patient to a doctor during the consultation process.

Groupthink
Psychological phenomenon describing the tendency of members of a group to conform to consensus opinion.

Hacking
Use of computers to obtain unauthorized access to data.

Halo effect
A good impression made by one characteristic, such as attractiveness, extended without justification to other areas, such as honesty.

Health anxiety
Excessive or obsessive concern about one's health; preoccupation with perceived or potential illness.

Human-computer interaction (HCI)
The study of how people interact with computers and how successfully computers are designed for users.

Human trafficking
The forced movement of a person for the purpose of exploitation.

Hyperpersonal interaction
In online environments, where users may be selective and edit their communications, idealized self-images are constructed and received, and intimacy is quickly established.

Iatrogenic death (Iatrogenesis)
Inadvertent and preventable induction of disease or complications by a medical treatment that results in the death of the patient.

Identity deception
Deliberate promulgation of misinformation about one's identity.

Identity formation
Development of an individual's personality, beginning in childhood and usually maturing through adolescence.

Impression management
Presenting a favorable and/or enhanced public image of oneself so that others will form positive judgments.

Impulsive behavior
Characterized by the urge to act spontaneously without reflecting on an action and its consequences.

Instamacy
Slang for too-swiftly established feelings of intimacy or trust.

Intelligence amplification (IA)
The use of information technology to augment human intelligence.

Internet addictive behavior
Compulsive behavior resulting from escalating reliance on Internet services or the need to satisfy a craving for Internet-related activity, involving distress caused by its withdrawal. Also called *Internet use disorder, Internet addiction, problematic Internet use, dysfunctional Internet behavior, virtual addiction.*

Internet gaming disorder
Excessive participation in Internet-based games that leads to significant behavioral or mental dysfunction. Also called *Internet gaming addiction, compulsive Internet gaming, online gaming addiction.*

Internet service provider (ISP)
A company providing access to the Internet.

Internet shopping addiction
Compulsive, episodic purchasing of goods facilitated and exacerbated by the accessibility of items online. Also called *online shopping addiction, eBay addiction, compulsive shopping.*

Locard's exchange principle
Principle of forensic science that every contact with a person, place, or thing results in an exchange that can be traced and used as evidence.

Locus of control
The extent to which people believe they have power over events in their lives. People with an internal locus of control believe that they are responsible for their own success or failure; those with an external locus of control believe that external forces, such as luck, determine their outcomes.

Longitudinal study
A research method in which data is gathered by observation of the same subjects over an extended period of time.

Looking-glass theory
A metaphor describing how an individual's identity can develop in response to feedback about how others see them.

Maladaptive behavior
Behavior that interferes with the activities of daily life or that is inappropriate in a given setting.

Malware (malicious software, Trojans, keyloggers, ransomware, spyware)
Software designed to infiltrate and disrupt or damage a computer or computer network.

Man-computer symbiosis
Potentially interdependent relationship between human beings and machines.

Mindfulness
Awareness of the present moment and acceptance of the current state of being.

Minimization of status and authority online
Construct that authority figures wield less influence online, without the real-world environment or trappings that reinforce status.

Mirror-image stimulation
A technique whereby an organism is confronted with its own reflection in a mirror.

Mobile phone addiction
Compulsive and excessive use of mobile/cell phones.

Mule (money mule, storage mule, smurfer)
Carrier of illicit items on behalf of another person; a money mule receives and transfers cash for laundering.

Multiplayer game
Role-playing video game conducted online so that a very large number of people can participate simultaneously and interact. Also called *massive multiplayer online role playing game (MMORPG).*

Munchausen by Internet
Feigning, exaggerating, or self-inducing illness to command attention, and carrying out this deception online.

Narcissism
Excessive admiration of oneself and/or one's own appearance, often combined with self-aggrandizement and an extreme craving for admiration.

Narcissistic personality disorder
Personality disorder in which an individual's inflated sense of self-importance, deep need for admiration, and lack of empathy for others often masks hypersensitivity to criticism.

Narcissistic Personality Inventory
A measure designed to explore narcissistic behavior and sentiments.

Neurotransmitters
Brain chemicals that relay signals between nerve cells and communicate information throughout the brain and body.

Obsessive-compulsive disorder (OCD)
Anxiety disorder characterized by obsessions and compulsions and unreasonable thoughts and/or fears that lead to repetitive and/or unnecessary behaviors.

Online disinhibition effect (ODE)
The tendency to self-disclose, to say and do things in cyberspace that would not ordinarily be said or done in the real world.

Online syndication
Use of the Internet to find and associate with other like-minded individuals, to normalize and socialize underlying tendencies, and to combine in a joint effort.

Open privacy
Refers to contemporary understanding of privacy, particularly among youth.

Paraphilia
A preference for atypical and unusual sexual practices.

Parental controls
Features of a digital service permitting adults to limit or filter content unsuitable for viewing by children.

PC bangs
Large establishments, chiefly in Asia, offering the use of computers with Internet connections to gamers for a fee. Also called *Internet café gaming centers, local area network (LAN) centers.*

Phone phreaking
Exploring a telephone system or manipulating it to use services without payment.

Photobomb
Unexpected intrusion into the field of view as a photograph is taken.

Piracy
Unauthorized access, use, copying, or distribution of another's work.

Planned behavior
In psychology, a theory that links beliefs and behavior.

Primacy effect
An attractive trait or feature standing out in the first impression of an individual that may have the effect of overshadowing other features.

Primal fear
An anxious feeling—a primitive fear that has some basic survival value.

Privacy paradox
The desire of an individual to maintain privacy in conflict with the practice of sharing sensitive personal information online.

Pro-techno-social initiatives
Technological initiatives aimed at resolving technology-facilitated social and behavioral problems.

Psychopathy
A mental disorder characterized by enduring antisocial behavior, along with diminished empathy and remorse.

Psychosomatic effect
Illness or disorder caused or exacerbated by psychological or emotional factors.

Remote Access Trojan (RAT)
Software program introduced to a victim's computer for malicious purposes, often compromising administrative control.

Revenge porn
The public sharing of indecent or explicit images of a person without their consent.

Risky-shift phenomenon
An individual's tendency toward riskier behavior as a result of the influence of a group.

Robotics
The science or study of the technology associated with the design, manufacture, theory, and application of robots.

Routine activity theory (RAT)
Environmental or location-based theory of crime that explains how opportunities for crime are produced through day-to-day activities, and determines the likelihood of when and where criminal events may occur.

Sadism
Deriving gratification, especially sexual pleasure, from inflicting pain and/or humiliation upon another.

Scareware
False security warnings encouraging a victim to install damaging software.

Self-actualization
The desire for self-fulfillment to maximize one's potential.

Self-concept
A general term used to refer to how someone thinks about themselves; an individual's belief about themselves, including the person's attributes and who and what the self is.

Selfie
A photograph taken of oneself, typically with a smartphone or webcam and shared via social media.

Sensorimotor development
According to Piaget's Theory of Cognitive Development, a stage of development from birth to about twenty-four months of age when a baby uses its senses and motor skills to learn about its own behavior and reacts to different stimuli, such as movement and emotion.

Sex addiction
Compulsive and escalating participation in sexual activities to the extent that other activities and interactions are negatively affected. Also called hypersexual disorder, compulsive sex behavior.

Sextortion
A form of blackmail and sexual exploitation that employs nonphysical forms of coercion by threatening to release sexual images or information to extort sexual favors or money from the victim.

Sexts
The sending of sexually explicit messages or photos by using a cellphone or other device.

Signaling theory
Transmission of information from a sender to a receiver understood in terms of evolutionary biology.

Situation 21
One of the worst-case scenarios rejected in security preparations for the 1972 Olympic Games as unlikely but that actually came to pass; now an axiom that hoping for the best will not prevent the worst from occurring.

Socially engineered attack
Strategy of attack inducing human error, often tricking or manipulating individuals into breaching normal security protocols.

Stranger on the train syndrome
Metaphor describing an individual's willingness to disclose personal information or have an intimate discussion with a stranger or passing acquaintance.

Suggestible
Easily influenced by the opinion of another.

Surface Web
Any part of the World Wide Web that is readily available to the general public and searchable with standard Internet search engines.

Techno-behavioral effect
The impact of technology on human behavior.

Techno-social effect
The impact of technology on society.

Technosomatic effect
Psychosomatic symptoms amplified by online interactions.

Tech rage

Extreme anger caused by frustration with the operation of a technological device or system, sometimes accompanied by physical or verbal abuse directed at the device.

Time-distortion effect

Losing track of time while immersed in an online environment.

Tor (the onion router)

Free software for enabling anonymous use of the Internet.

TQ (Technology Quotient)

Suggested new metric or scale to identify, assess, and measure technological abilities.

Transdisciplinary research

Investigators from different disciplines working jointly to create integrated solutions to common problems.

Transference

In psychoanalysis, a form of displacement—for example assigning characteristics of a person from one's past, or projecting thoughts and wishes associated with that person, to a person in the present.

Triple A Engine

Referring to factors that explain the power and attraction of the Internet for sexual pursuits; anonymity, accessibility, and affordability.

Troll

A person deliberately posting malicious or inflammatory messages with the intent to provoke a negative response.

Ubiquitous victimology

Term describing the wide pool of potential high-risk crime victims owing to global proliferation of technology.

Uncanny Valley

The feeling of unease aroused by computer-generated figures or robots closely resembling human beings.

Unnatural game design

Online games containing incentives to continue play beyond points of normal physical or mental fatigue.

Visual acuity

Commonly refers to the clarity of vision.

Webcam sex tourism

Sex offenders paying to direct and view live-streaming video footage of children in another country performing sexual acts in front of a webcam.

Withdrawal

Distress caused to someone with a psychological or physical dependency when the addictive material is withdrawn.

NOTES

Below you'll find a selection of references that informed this book. For additional references and material, please go to maryaiken.com.

Prologue: When Humans and Technology Collide

6. **"God is in the details":** The idiom "The devil is in the details" means that mistakes are usually made in the small details of a project or exercise. Usually it is a caution to pay attention in order to avoid failure. An older and more common phrase, "God is in the details," means that attention paid to small things has big rewards; in other words, details are important.
7. ***cyber juvenile delinquency* (hacking):** M. P. Aiken (2016), "Not Kidding," *Freud's The Brewery Journal: Cybercrime* 6: 48–51.
9. **the number of people with access to the Internet:** itu.int, May 26, 2015.
9. **The number of hours people spend on mobile phones:** "Smartphones: So Many Apps, So Much Time," Nielsen.com/us, July 1, 2014.
9. **checked their devices more than fifteen hundred times:** "How Often Do YOU Look at Your Phone?," *MailOnline,* October 7, 2014.
9. **there are several apps that will count:** The advent of the "checkyapp" (checkyapp.com), intended to help raise mobile phone users' awareness of their habits, was reported by S. Perez, "How Many Times a Day Do You Check Your Phone? Checky Will Tell You," TechCrunch.com, September 15, 2014.
16. **"The Good, the Bad, and the Unknown":** J. S. Radesky, J. Schumacher, and

B. Zuckerman (2014), "Mobile and Interactive Media Use by Young Children: The Good, the Bad, and the Unknown," *Pediatrics* 135(1): 1–3.

Chapter 1: The Normalization of a Fetish

22. **the *online disinhibition effect*:** J. Suler (2004), "The Online Disinhibition Effect," *Cyberpsychology and Behavior* 7(3): 321–26; and J. Suler (2005), "The Online Disinhibition Effect," *International Journal of Applied Psychoanalytic Studies* 2(2): 184–88. For Suler's work in general, see his regularly revised online book *The Psychology of Cyberspace*, at rider.edu.
22. ***online escalation*:** R. W. White and E. Horvitz (2009), "Cyberchondria" *ACM Transactions on Information Systems* 27(4), Article No. 23. Also see M. Aiken, G. Kirwan, M. Berry, and C. A. O'Boyle (2012), "The Age of Cyberchondria," *Royal College of Surgeons in Ireland Student Medical Journal* 5(1): 71–74.
23. **Cyberstalking:** K. Baum, S. Catalano, M. Rand, and K. Rose (2009), *Stalking Victimization in the United States*, U.S. Department of Justice, pp. 1–16, victimsofcrime.org.
23. **he announced he was running for political office:** Jordan Haskins campaign website: jordanhaskinsfor95thdistrictstaterepresentative.yolasite.com.
23. **"I've found my niche":** "Felony Convictions Linked to Sexual Fetish 'Haunt Me,'" *Mlive.com*, June 27, 2014.
24. **the most common fetishes:** M. D. Griffiths, "Survival of the Fetish: A Brief Overview of Bizarre Sexual Behaviours," psychologytoday.com, January 7, 2014.
25. **"When now I announce that the fetish":** J. Strachey (trans.) (1964), *The Complete Psychological Works of Sigmund Freud*, vol. XXI (London: The Hogarth Press and the Institute of Psychoanalysis), 152–53.
25. **a boy may associate arousal:** R. Crooks and K. Baur (2011), *Our Sexuality,* 11th ed. (Belmont, Calif.: Wadsworth), 499.
26. **an anticipation of a reward:** S. A. McLeod (2007/2013), "Pavlov's Dogs," simplypsychology.org.
26. **the "'vroom' of the engine":** "Growing Fetish Trend: Pedal-Pumping, Revving and Cranking," *The Independent* (online), March 29, 2010.
26. **In one classical conditioning experiment:** S. Rachman and R. J. Hodgson (1968), "Experimentally-Induced 'Sexual Fetishism': Replication and Development," *Psychological Record* 18: 25–27.
29. **To be diagnosed with a paraphilia disorder:** American Psychiatric Association (2013), *Diagnostic and Statistical Manual of Mental Disorders, DSM-5* (Washington, DC: American Psychiatric Publishing), 685–705.
30. **responding sexually to sadomasochistic narratives:** A. C. Kinsey, W. B.

Pomeroy, C. E. Martin, and P. H. Gebhard (1953), *Sexual Behavior in the Human Female* (Philadelphia: W. B. Saunders Company), 676–78.

30. **sexual gratification from inflicting pain:** M. Hunt (1974), *Sexual Behavior in the 1970s* (New York: Playboy Press).
31. **sadomasochistic sexual fantasies:** R. B. Krueger (2010), "The DSM Diagnostic Criteria for Sexual Masochism," *Archives of Sexual Behavior* 39(2): 346–56.
31. **reported to have rape fantasies:** J. W. Critelli and J. M. Bivona (2008), "Women's Erotic Rape Fantasies: An Evaluation of Theory and Research," *The Journal of Sex Research* 45(1): 57–70.
31. **tying up their partner:** W. B. Arndt, Jr., J. C. Foehl, and F. E. Good (1985), "Specific Sexual Fantasy Themes: A Multidimensional Study," *Journal of Personality and Social Psychology* 48(2): 472–80.
32. **disturbing aspect of *Fifty Shades:*** On her personal webpage, eljamesauthor.com, *Fifty Shades* author E. L. James describes her trilogy as "adult romance."
32. **bondage and domination imagery:** P. E. Dietz and B. Evans (1982), "Pornographic Imagery and Prevalence of Paraphilia," *American Journal of Psychiatry* 139(11): 1493–95.
32. **Internet searches for BDSM porn:** "'Fifty Shades of Grey' Effect: BDSM More Popular Than Ever, Especially with New Yorkers," booksnreview.com, August 30, 2012. See also M. Haber, "A Hush-Hush Topic No More," NYTimes.com, February 27, 2013.
32. **Membership in FetLife:** FetLife.com.
33. **individuals who practice sadomasochism:** R. B. Krueger (2010), "The DSM Diagnostic Criteria for Sexual Masochism," *Archives of Sexual Behavior* 39(2): 346–56.
33. **"sociological and social psychological studies":** T. S. Weinberg (2006), "Sadomasochism and the Social Sciences: A Review of the Sociological and Social Psychological Literature," *Journal of Homosexuality* 50(2): 37.
35. **Several unusual things were found:** D. McDonald and N. Anderson, "Graham Dwyer Trial: Latex Bodysuit Found in Elaine O'Hara's Flat, Court Hears," Irish *Independent,* January 1, 2015.
35. **Elaine's emotional age:** C. Gleeson, "Father's Partner Tells of Concerns over Elaine O'Hara Self-Harming," *The Irish Times*, January 26, 2015.
35. **she had asked him to kill her:** C. ÓFátharta and N. Dwyer, "Elaine's Father: Architect Refused Request to Kill Her," *Irish Examiner,* January 24, 2015.
35. **her remains were found in the underbrush:** E. Edwards, "Graham Dwyer Trial: Knives, Mobile Phone and Handcuffs Among Items in Reservoir," *Irish Times,* January 29, 2015.

35. **"My urge to rape, stab or kill":** P. Flanagan and N. Reid, "Murder Trial Told Graham Dwyer Sent a Text to Victim Allegedly Saying: 'My Urge to Rape, Stab or Kill Is Huge,'" *Irish Mirror,* January 22, 2015.
35. **O'Hara left behind a notebook:** D. McDonald and N. Anderson, "Graham Dwyer Trial: Latex Bodysuit Found in Elaine O'Hara's Flat, Court Hears," Irish *Independent,* January 1, 2015.
36. **"Technology is a killer":** C. Cromie, S. Stack, and B. Hutton, "Graham Dwyer Guilty: Sadist Architect Stabbed Dublin Woman Elaine O'Hara to Death During Sex," *Belfast Telegraph,* March 27, 2015.
38. **now bans images that depict abusive:** H. Saul, "UK Porn Legislation: What Is Now Banned Under New Government Laws," *The Independent,* December 2, 2014.
40. **he befriended female strangers online:** R. Weiner, "Anthony Weiner Details How Many Women He's Had Online Relationships With," *The Washington Post,* July 25, 2013.
40. **study at Ohio State University:** R. Nauert, "Posting of Selfies May Suggest Personality Issues," psychcentral.com, January 7, 2015.
40. **men who post a lot of selfies:** J. Fox and M. C. Rooney (2015), "The Dark Triad and Trait Self-Objectification as Predictors of Men's Use and Self-Presentation Behaviors on Social Networking Sites," *Personality and Individual Differences* 76: 161–65.
42. **A bizarre case in the U.K.:** E. Griffiths, "Warning over new 'cyber flashing' crime after woman's iPhone is bombarded with explicit images on train," *Mirror,* August 13, 2015.
43. **"I want to again say":** R. Weiner, "Anthony Weiner Acknowledges More Explicit Texts," *The Washington Post,* July 23, 2013.
43. **"Maybe if the Internet didn't exist":** M. Sella, "The Year of Living Carlos Dangerously," *GQ,* gq.com, October 16, 2013.
43. **Instead, he lost the mayoral primary:** "New York City Primary Results," NYTimes.com, September 16, 2013.
43. **he whipped out his middle finger:** The moment was captured in photographs by @KateRoseMe and @LindseyChrist and posted on Twitter, where they are still available for viewing. See W. Hickey, "Photo: Anthony Weiner Ends His Campaign by Giving the Press the Finger," businessinsider.com, September 10, 2013.
43. ***voyeurism,* also known as *scopophilia:*** B. J. Sadock and V. A. Sadock (2008), *Kaplan & Sadock's Concise Textbook of Clinical Psychiatry,* 3rd ed. (Philadelphia: Lippincott Williams & Wilkins), 322.
43. **Rabbi Barry Freundel:** K. L. Alexander, S. Pulliam Bailey, and M. Boorstein, "D.C. Rabbi Pleads Guilty to Voyeurism Charges," *The Washington Post,* February 19, 2015.

44. **Jared James Abrahams:** K. Gander, "Miss Teen USA Webcam Hacker Jared James Abrahams Sentenced to 18 Months in Prison," *The Independent,* March 18, 2014.
44. **"I wasn't aware that somebody was watching":** "More Than 90 People Arrested in 'Creepware' Hacker Sting as Victim Miss Teen USA Describes 'Terror' at Being Watched Through Her Webcam for a YEAR," *MailOnline,* May 25, 2014.
45. **more than 70 million computers:** statistics sourced from Gartner, International Data Corporation, and published on statisticbrain.com, January 14, 2015.
45. **unknown fetishes like cranking:** "10 unusual fetishes with massive online followings," criminaljusticedegreesguide.com.

Chapter 2: Designed to Addict

46–47. **Alexandra Tobias:** News4Jax.com, February 1, 2011. Notably it has been reported that the day before the death, Tobias took a personality test on the Internet, which labeled her bipolar: D. Hunt, "Jacksonville Mom Who Killed Baby While Playing FarmVille Gets 50 Years," *The Florida Times-Union,* February 1, 2011.
47. **impulsivity:** C. G. Coutlee, C. S. Politzer, R. H. Hoyle, and S. A. Huettel (2014), "An Abbreviated Impulsiveness Scale Constructed Through Confirmatory Factor Analysis of the Barratt Impulsiveness Scale Version 11," *Archives of Scientific Psychology* 2: 2.
48. **a list of thirty simple statements:** J. H. Patton, M. S. Stanford, and E. S. Barratt (1995), "Factor Structure of the Barratt Impulsiveness Scale," *Journal of Clinical Psychology* 51(6): 768–774.
48. **response inhibition:** W. Ding et al. (2014), "Trait Impulsivity and Impaired Prefrontal Impulse Inhibition Function in Adolescents with Internet Gaming Addiction Revealed by a Go/No-Go fMRI Study," *Behavioral and Brain Functions* 10: 20.
48. **excessive Internet use:** F. Cao and L. Su (2007), "Internet Addiction Among Chinese Adolescents: Prevalence and Psychological Features," *Child: Care, Health and Development* 33(3): 275–81. See also G. J. Meerkerk, R. J. J. M. van den Eijnden, I. H. A. Franken, and H. F. L. Garretsen (2010), "Is Compulsive Internet Use Related to Sensitivity to Reward and Punishment, and Impulsivity?" *Computers in Human Behavior* 26(4): 729–35.
49. **study of pigeons:** D. A. Eckerman and R. N. Lanson (1969), "Variability of Response Location for Pigeons Responding Under Continuous Reinforcement, Intermittent Reinforcement, and Extinction," *Journal of the Experimental Analysis of Behavior* 12(1): 73–80.

50. **Dopamine:** P. M. Newton, "What Is Dopamine?: The Neurotransmitter's Role in the Brain and Behavior," psychologytoday.com, April 26, 2009.

50. **called "a heart-stopper":** M. Woolf, "That Irresistible Urge to Scratch: Lottery/Instants Fever," *The Independent,* April 15, 1995.

50. **In game design this is called "fun failure":** M. Breeze, "A Quiet Killer: Why Video Games Are So Addictive," TheNextWeb.com, January 12, 2013.

51. **"stimulates the release of loads of dopamine":** E. Ritvo, "Facebook and Your Brain: The Inside Dope on Facebook," psychologytoday.com, May 24, 2012.

51. **a brain response similar to the release of dopamine:** D. I. Tamir and J. P. Mitchell (2012), "Disclosing Information About the Self Is Intrinsically Rewarding," *Proceedings of the National Academy of Sciences of the United States of America* 109(21): 8038–43.

52. **"It is the mammalian motivational engine":** E. Yoffe, "Seeking: How the Brain Hard-Wires Us to Love Google, Twitter, and Texting. And Why That's Dangerous," Slate.com, August 12, 2009.

52. **what we think of as our "core-self":** K. Badt, "Depressed? Your 'SEEKING' System Might Not Be Working: A Conversation with Neuroscientist Jaak Panksepp," huffingtonpost.com, September 17, 2013.

53. **The Latin word *addictus:*** R. E. Cytowic, "Ambivalence in Addiction," psychologytoday.com, November 13, 2015.

54. **A 2015 study found that Americans:** L. Eadicicco, "Americans Check Their Phones 8 Billion Times a Day," Time.com, December 15, 2015.

54. **tests to take online:** behaviorhealth.bizcalcs.com.

56. **signaling theory in marketing:** B. Dunham (2011), "The Role for Signaling Theory and Receiver Psychology in Marketing," in G. Saad (ed.), *Evolutionary Psychology in the Business Sciences.* (Heidelberg, Germany: Springer), 225–56.

57. **apps created to help . . . break these patterns of behavior:** K. Montgomery, "To Solve Phone Addiction, App Shows How Many Times You Check Your Phone," Gawker.com, September 16, 2014.

59. **For 2017, the number of cellphone users:** Forecast of number of cellphone users worldwide (2013 to 2019) found at www.statista.com.

59. **dangers of being "too connected":** K. Young (2015), "What You Need to Know About Internet Addiction," tedxtalks.ted.com/video/What-You-Need-to-Know-About-Int.

60. **The cravings they described:** E. Aboujaoude et al. (2006), "Potential Markers for Problematic Internet Use: A Telephone Survey of 2,513 Adults," *CNS Spectrums* 11(10): 750–55.

60. **Korea . . . China:** See M. H. Hur (2006), "Demographic, Habitual, and Socioeconomic Determinants of Internet Addiction Disorder: an Empirical

Study of Korean Teenagers," *Cyberpsychology & Behavio*r 9(5): 514–25; J. J. Block (2008), "Issues for DSM-V: Internet Addiction," *American Journal of Psychiatry* 165 (3): 306–7.

60. **Italian adolescents:** L. Milani, D. Osualdella, and P. Di Blasio (2009), "Quality of Interpersonal Relationships and Problematic Internet Use in Adolescence," *CyberPsychology & Behavior* 12(6): 681–4.

60. **seven European countries:** A. Tsitsika et al. (2014), "Internet Addictive Behavior in Adolescence: A Cross-Sectional Study in Seven European Countries," *CyberPsychology, Behavior & Social Networking* 17(8): 528–35.

61. **"overusing, abusing, or misusing their devices":** K. Wallace, "10 Signs You Might Be Addicted to Your Smartphone," CNN.com, November 25, 2014.

63. **"the need to seek out variety":** E. Hartney, "Is Compulsive Shopping Really an Addiction?," verywell.com, April 5, 2016.

64. **active users of eBay:** statista.com

64. **"I feel a literal rush":** momma*jess, "Retail Therapy Syndrome: Shopping Addiction 101," eBay.com, March 16, 2007.

64. **"In more serious cases":** K. Young, netaddiction.com/ebay.

66. **"Chris lived for his Xbox":** R. Twomey, "Xbox Addict, 20, Killed by Blood Clot After 12-Hour Gaming Sessions," *Daily Mail*, July 30, 2011 (updated January 6, 2016).

67. **DVT is caused by inactivity:** M. G. Beckman, W. C. Hooper, S. E. Critchley, and T. L. Ortel (2010), "Venous Thromboembolism: A Public Health Concern," *American Journal of Preventive Medicine* 38(suppl. 4): S495–S501.

67. **"gamer's thrombosis":** B. Gholipour, "Gamer's Thrombosis: How Playing Too Long Could Be Deadly," LiveScience.com, December 10, 2013.

68. **A number of fatalities:** A. Rudd, "Diablo Death: Teenager Dies After Playing Video Game for 40 Hours Without Eating or Sleeping," *Mirror,* July 18, 2012.

68. **professional gaming is now a multimillion-dollar industry:** S. Y. Hwang, "South Korea's Game Addiction Law Could Treat Games Like Drugs and Alcohol," CNET.com, June 23, 2014.

68. **A spike of concern was provoked:** M. Tran, "Girl Starved to Death While Parents Raised Virtual Child in Online Game," *The Guardian,* March 5, 2010.

69. **The results showed that their cravings:** C. Ko et al. (2009), "Brain activities associated with gaming urge of online gaming addiction," *Journal of Psychiatric Research* 43(7): 739–47.

70. **Internet Gaming Disorder:** Internet Gaming Disorder is a "Condition for Further Study" in the DSM-5: American Psychiatric Association (2013), *Diagnostic and Statistical Manual of Mental Disorders 5th edition* (Arlington, VA: American Psychiatric Publishing).

70. **The APA suggested that these four:** A. Weinstein and M. Lejoyeux (2010), "Internet Addiction or Excessive Internet Use," *American Journal of Drug and Alcohol Abuse* 36(5): 277–83.
70. **In South Korea, the government:** "Gaming Addiction Is Real and Growing Problem," *The Korea Herald,* November 13, 2013.
71. **Taiwan:** M. Locker, "This Place Just Made It Illegal to Give Kids Too Much Screen Time," Time.com, January 26, 2015.
71. **"we see addicted gamers":** K. Young, "Should video games be considered a collegiate sport? I say No . . ." netaddictionrecovery.blogspot.com, November 10, 2014.
72. **Stanford Marshmallow Experiments:** The original Stanford experiment involved pretzels and animal cookies, but no marshmallows: W. Mischel and E. B. Ebbesen (1970), "Attention in Delay of Gratification," *Journal of Personality and Social Psychology* 16(2): 329–37. The marshmallows were introduced in a 1972 follow-up study: W. Mischel, E. B. Ebbesen, and A. Raskoff Zeiss (1972), "Cognitive and Attentional Mechanisms in Delay of Gratification," *Journal of Personality and Social Psychology* 21(2): 204–18.

    On educational achievements, see Y. Shoda, W. Mischel, and P. K. Peake (1990), "Predicting Adolescent Cognitive and Self-Regulatory Competencies from Preschool Delay of Gratification: Identifying Diagnostic Conditions," *Developmental Psychology* 26(6): 978–86.

    On BMI, see T. R. Schlam, N. L. Wilson, Y. Shoda, W. Mischel, and O. Ayduk (2013), "Preschoolers' Delay of Gratification Predicts Their Body Mass 30 Years Later," *The Journal of Pediatrics* 162(1): 90–93.
72. **children and teens diagnosed with some behavioral disorders:** L. Davis (n.d.), "Risk of Internet Addiction Higher in Teens with ADHD and Depression," video-game-addiction.org.
74. **compulsive behaviors are "produced" by different games:** M. Breeze, "A Quiet Killer: Why Video Games Are so Addictive," TheNextWeb.com, January 12, 2013.
78. **(As many as 41 percent):** Z. Hussain and M. D. Griffiths (2009), "Excessive Use of Massively Multi-Player Online Role-Playing Games: A Pilot Study," *International Journal of Mental Health and Addiction* 7(4): 563–71.
79. **depression and "low locus of control":** E. Andreou and H. Svoli (2013), "The Association Between Internet User Characteristics and Dimensions of Internet Addiction Among Greek Adolescents," *International Journal of Mental Health and Addiction* 11(2): 139–48.
79. **the "power of positive thinking":** Norman Vincent Peale's famous book continues to resonate over half a century after it was first published: N. V. Peale (1952), *The Power of Positive Thinking* (Upper Saddle River, N.J.: Prentice-Hall).

79. **negative social, behavioral, and health consequences:** W. Ding et al. (2014), "Trait Impulsivity and Impaired Prefrontal Impulse Inhibition Function in Adolescents with Internet Gaming Addiction Revealed by a Go/No-Go fMRI Study," *Behavioral and Brain Functions* 10: 20. See also A. Weinstein and M. Lejoyeux (2010), "Internet Addiction or Excessive Internet Use," *American Journal of Drug and Alcohol Abuse* 36(5): 277–83.
80. **"[T]hey promise more than they deliver":** C. Peterson, S. F. Maier, M. E. P. Seligman (1993), *Learned Helplessness: A Theory for the Age of Personal Control* (New York: Oxford University Press), 307–8.
81. **In 2015, the cellphone pornography business:** internetsafety101.org/mobile statistics.htm.
82. **"Hypersexual Disorder":** M. P. Kafka (2010), "Hypersexual Disorder: A Proposed Diagnosis for DSM-V," *Archives of Sexual Behavior* 39: 377–400.
82. **the APA chose to disregard:** R. Weiss, "New Research Supports Sexual Addiction as a Legitimate Diagnosis," rehabs.com, August 18, 2014.
83. **"the Triple A Engine":** A. Cooper (1998), "Sexuality and the Internet: Surfing into the New Millennium," *Cyberpsychology & Behavior* 1(2): 187–93.
84. **Both gaming and porn:** P. Zimbardo (2011), "The Demise of Guys?," TED .com.
85. **treatment of post-traumatic stress disorder:** M. P. Aiken and M. J. Berry (2015), "Posttraumatic Stress Disorder: Possibilities for Olfaction and Virtual Reality Exposure Therapy," *Virtual Reality* 19(2): 95–109.
86. ***Maladaptive behavior:*** D. Busse and I. S. Yim (2013), in M. D. Gellman and J. R. Turner (eds.), *Encyclopedia of Behavioral Medicine* (New York: Springer), 1187–88.

Chapter 3: Cyber Babies

90. **studying baby rhesus monkeys:** The Adoption History Project (n.d.), "Harry F. Harlow, Monkey Love Experiments," pages.uoregon.edu/adoption/studies/HarlowMLE.htm.
95. **Synapse formation for key developmental functions:** B. Worthen, "What Happens When Toddlers Zone Out with an iPad," *The Wall Street Journal,* May 22, 2012.
96. **During the first two months of life:** M. M. Haith (1980), *Rules That Babies Look By: The Organization of Newborn Visual Activity* (Hillsdale, N.J.: Lawrence Erlbaum Associates).
97. **Eventually several studies backed up:** J. S. DeLoache et al. (2010), "Do Babies Learn from Baby Media?," *Psychological Science* 21(11): 1570–74.
    See also R. A. Richert, M. B. Robb, J. G. Fender, and E. Wartella (2010),

"Word Learning from Baby Videos," *Archives of Pediatrics and Adolescent Medicine* 164(5).

97. **study claimed that these videos could delay speech:** F. J. Zimmerman, D. A. Christakis, and A. N. Meltzoff (2007), "Associations Between Media Viewing and Language Development in Children Under Age 2 Years," *The Journal of Pediatrics* 151(4): 364–68.

97. **the videos had minimal to no impact on infant development:** J. Martin, "UW Battle over Baby Einstein Settled, Maybe," *The Seattle Times,* June 30, 2011.

98. **more effective . . . quieter shows, calmer formats:** K. Morelli, "The I-Baby: A Baby's Brain on Technology," scienceandsensibility.org, April 11, 2013.

98. **young children learn through gestures:** M. L. Rowe and S. Goldin-Meadow (2009), "Early Gesture *Selectively* Predicts Later Language Learning," *Developmental Science* 12(1): 182–87.

98. **Quickly dubbed the "worst toy of the year":** "Fisher Price Under Pressure to Pull the Plug on New iPad Baby Bounce Seat Aimed at Newborns over Claims It Is 'Unhealthy,'" *Daily Mail,* December 10, 2013.

99. **Let's start with the fact:** In 2011 the World Health Organization's International Agency for Research on Cancer (IARC) classified the radio frequency electromagnetic fields of cellphones as a "Group 2B risk," i.e., as "possibly carcinogenic to humans": World Health Organization (2014), *Electromagnetic Fields and Public Health: Mobile Phones,* who.int.mediacentre/factsheets/fs193/en/, reviewed October 2014. More recently, in 2013, a review of earlier studies and new research recommended that radio frequency exposure should be reclassified as a "Group 2A risk," i.e., a "probable human carcinogen": D. L. Davis, S. Kesari, C. L. Soskolne, A. B. Miller, and Y. Stein (2013), "Swedish Review Strengthens Grounds for Concluding That Radiation from Cellular and Cordless Phones Is a Probable Human Carcinogen," *Pathophysiology* 20: 123–29.

99. **kittens . . . visual experiences at birth:** C. Blakemore and G. F. Cooper (1970), "Development of the Brain Depends on the Visual Environment," *Nature* 228: 477–78; see YouTube.com (content may be disturbing for some). See also D. H. Hubel and T. N. Wiesel (1970), "The Period of Susceptibility to the Physiological Effects of Unilateral Eye Closure in Kittens," *Journal of Physiology* 206(2): 419–36.

100. **But let's start with some things that are proven:** B. Keim, "It's Official: To Protect Baby's Brain, Turn Off TV," *Wired,* October 18, 2011.

100. **recommended against screen use:** American Academy of Pediatrics (2011), "Media Use by Children Younger Than 2 Years," *Pediatrics* 128(5): 1040–45.

101. **review paper, "Cyber Babies":** C. Haughton, M. P. Aiken, and C. Cheevers

(2015), "Cyber Babies: The Impact of Emerging Technology on the Developing Infant," *Psychology Research,* 5(9): 504–18. DOI:10.17265/2159-5542/2015.09.002.

102. **tablets and mobile phones . . . to placate and pacify young children:** L. McDonald and L. Colgan, "Experts Claim iPhones and iPads Are Now Being Used as 'Virtual Childminders' for Kids as Young as Two," Evoke.ie, February 2, 2015.

104. **"the Skinner box":** N. Joyce and C. Faye (2010), "Skinner Air Crib," *Observer* 23(7).

106. **Moving to Learn, who recommends:** C. Rowan, "Ten Reasons Why Handheld Devices Should Be Banned for Children Under the Age of 12," movingtolearn.ca, February 24, 2014.

107. **British teachers report:** R. Ratcliffe, "Children Can Swipe a Screen but Can't Use Toy Building Blocks, Teachers Warn," *The Guardian,* April 15, 2014.

108. **"The Birth of a Word":** D. Roy, TED Talk: ted.com, March 2011.

108. **Screen use can cause sleep disturbances:** On screen use before bed for adults: A.-M. Chang, D. Aeschbach, J. F. Duffy, and C. A. Czeisler (2015), "Evening Use of Light-Emitting eReaders Negatively Affects Sleep, Circadian Timing, and Next-Morning Alertness," *Proceedings of the National Academy of Sciences* 112(4): 1232–37.

On screen use before bed for children: L. S. Foley, R. Maddison, Y. Jiang, S. Marsh, T. Olds, and K. Ridley (2013), "Presleep Activities and Time of Sleep Onset in Children," *BMC Pediatrics* 131(2): 276–82.

108. **true of all "light-emitting devices":** E. A. Vandewater, V. J. Rideout, E. A. Wartella, X. Huang, J. H. Lee, and M. Shim (2007), "Digital Childhood: Electronic Media and Technology Use Among Infants, Toddlers, and Preschoolers," *Pediatrics* 119(5): e1006–15.

109. **"Learning about solitude":** S. Newman, "Are Screens 'Drugging' Your Child's Brain?," psychologytoday.com, September 2, 2014.

109. **In one study involving 2,463 children:** S. Shur-Fen Gau (2006), "Prevalence of Sleep Problems and Their Association with Inattention/Hyperactivity Among Children Aged 6–15 in Taiwan," *Journal of Sleep Research* 15(4): 403–14.

110. **connection between the rise of ADHD and screen use:** R. A. Friedman, "A Natural Fix for A.D.H.D.," *The New York Times,* October 31, 2014.

110. **Is it a coincidence:** Centers for Disease Control and Prevention (2015), "Attention-Deficit/Hyperactivity Disorder (ADHD): Data & Statistics." cdc.gov. For the full report, see S. N. Visser et al. (2014), "Trends in the Parent-Report of Health Care Provider–Diagnosed and Medicated Attention-Deficit/Hyperactivity Disorder: United States, 2003–2011," *Journal of the American Academy of Child and Adolescent Psychiatry* 53(1): 34–46.e2.

110. **The number of young people being treated with medication for ADHD:**

R. A. Friedman, "A Natural Fix for A.D.H.D.," *The New York Times,* October 31, 2014. But see the statistics in S. H. Zuvekas and B. Vitiello (2012), "Stimulant Medication Use in Children: A 12-Year Perspective," *The American Journal of Psychiatry* 169(2): 160–66, especially Table 2.

111. ***orienting response*****:** F. J. Zimmerman and D. A. Christakis (2007), "Associations Between Content Types of Early Media Exposure and Subsequent Attentional Problems," *Pediatrics* 120(5): 986–92.

111. **"There is no eye contact":** L. M. Oestreicher, "Re: The debate over digital technology and young people," bmj.com, August 25, 2015.

111. **television . . . attention problems:** D. A. Christakis, F. J. Zimmerman, D. L. DiGiuseppe, C. A. McCarty (2004), "Early Television Exposure and Subsequent Attentional Problems in Children," *Pediatrics* 113(4): 708–13; L. S. Pagani, C. Fitzpatrick, T. A. Barnett, and E. Dubow (2010), "Prospective Associations Between Early Childhood Television Exposure and Academic, Psychosocial, and Physical Well-being by Middle Childhood," *Archives of Pediatrics and Adolescent Medicine* 164(5): 425–31.

112. **"The decorated classroom led to greater time off task":** A. V. Fisher, K. E. Godwin, and H. Seltman (2014), "Visual Environment, Attention Allocation, and Learning in Young Children: When Too Much of a Good Thing May Be Bad," *Psychological Science* 25(7): 1362–70.

114. **Sherry Turkle describes a parent:** S. Turkle (2012), *Alone Together: Why We Expect More from Technology and Less from Each Other* (New York: Basic Books), 294, 266. See also Sherry Turkle, "Connected, but Alone?," TED Talk: ted.com, February 2012.

116. **half of all children under two:** kidshealth.org, January 2015.

116. **Beyond the Screens:** plus.google.com (2016).

116. ***Watching television can slow*****:** Associated Press, "France Bans Broadcast of TV Shows for Babies," Today.com, August 20, 2008.

116. **Taiwan has taken a giant:** J. W. Simons, "Why Taiwan Is Right to Ban iPads for Kids," CNN, February 4, 2015.

116. **Canada and Australia have similar recommendations:** Australian Government Department of Health and Ageing (2010), *Move and Play Every Day: National Physical Activity Recommendations for Children 0–5 Years.* See also Canadian Paediatric Society (2003), "Impact of Media Use on Children and Youth," *Paediatrics and Child Health* 8(5): 301–6.

Chapter 4: Frankenstein and the Little Girl

120. **"capitalised blood of children:":** K. Marx (1887), *Capital* vol. I, ed. F. Engels; S. Moore and E. Aveling (English trans.) republished 1959 (Moscow: Foreign Languages Publishing House), 756.

120. **the United Nations International Children's Emergency Fund:** UNICEF was created by the United Nations on December 11, 1946, to provide emergency food, clothing and healthcare to children in Europe and China after World War II. Since 1950 its mandate has been to address the needs of women and children in all developing countries.

121. **(84 percent) of U.S. children and teenagers have access to the Internet:** American Academy of Pediatrics (2013), "Children, Adolescents, and the Media," *Pediatrics* 132(5): 958.

121. **And as a marketing study shows:** "Survey: Majority of 'tweeners' now have cell phones . . ." nclnet.org, July 2012.

121. **most American children get their first cellphone:** "Study Finds Average Age of Kids When They Get Their First Cell Phone Is Six," abc13.com /technology, April 7, 2015.

122. **refined small-motor skills and enhanced hand-eye coordination:** D. Gozli, D. Bavelier, and J. Pratt (2014), "The Effect of Action Video Game Playing on Sensorimotor Learning: Evidence from a Movement Tracking Task," *Human Movement Science* 38: 152–62.

122. **a positive relation between texting and literacy:** B. Plester and C. Wood (2009), "Exploring Relationships Between Traditional and New Media Literacies: British Preteen Texters at School," *Journal of Computer-Mediated Communication* 14: 1108–1129. See also B. Plester, C. Wood, and P. Joshi (2009), "Exploring the Relationship Between Children's Knowledge of Text Message Abbreviations and School Literacy Outcomes," *British Journal of Developmental Psychology* 27(1): 145–61.

122. **if your children are texting a lot:** G. Paton, "Text Messaging 'Improves Children's Spelling Skills,'" telegraph.co.uk, January 20, 2011. Pre-test spelling performance was also measured, making the slight improvements after the relatively brief ten-week texting experiment more notable: C. Wood, E. Jackson, L. Hart, B. Plester, and L. Wilde (2011), "The Effect of Text Messaging on 9- and 10-Year-Old Children's Reading, Spelling and Phonological Processing Skills," *Journal of Computer Assisted Learning* 27: 28–36.

122. **"Some [children] cannot even talk":** E. Harding, "How Parents Who Use iPads to 'Pacify' Their Children Are Impeding Their Speech Development with Some Starting School at Five 'Unable to Talk' as a Result," dailymail .co.uk, January 31, 2015.

122. **mobile phones had become "virtual" babysitters:** R. Horan, "Psychologist: iPads are the New Childminders," 98fm.com, January 31, 2015.

124. **EU Kids Online:** S. Livingstone, L. Haddon, A. Görzig, and K. Ólafsson (2012), *EU Kids Online II: Final Report;* and (2014) *EU Kids Online: Findings. Methods. Recommendations.* (London: London School of Economics and Political Science), lse.ac.uk.

124. **United States:** C. K. Blackwell, A. R. Lauricella, A. Conway, E. Wartella (2014), "Children and the Internet: Developmental Implications of Web Site Preferences Among 8- to 12-Year-Old Children," *Journal of Broadcasting and Electronic Media* 58(1): 1–20.

125. **Four out of ten gave a false age:** K. Donnelly, "Four Out of 10 Children on Social Media Use False Age," Independent.ie, September 3, 2014.

125. **Twenty million minors use Facebook:** "That Facebook Friend Might Be 10 Years Old, and Other Troubling News," ConsumerReports.org, June 2011. Notably, Facebook's founder, Mark Zuckerberg, would like to see the minimum age requirement removed completely, to enable his company to begin educating children about the Internet from a much younger age: E. Protalinski, "Mark Zuckerberg: Facebook Minimum Age Limit Should Be Removed," ZDNet, May 20, 2011.

125. **"a majority of parents . . . unconcerned":** L. Magid, "Survey: 7.5M Facebook Users Below Minimum Age," CNET.com, May 10, 2011.

125. **Setting the minimum age:** The Children's Online Privacy Protection Act (COPPA) of 1998 (15 USC 6501-6506) requires a website or online service to obtain verifiable parental consent before it can make use of the data of those under the age of thirteen.

127. **the story of Sarah Lynn Butler:** Y. Young, "Online Teasing Leads to Teen's Suicide," Kait8.com, November 24, 2009.

127. **Let's look at how the numbers work on Facebook:** "Average Number of Facebook Friends of Users in the United States as of February 2014, by Age Group," statista.com.

128. ***Dunbar's number:*** R. A. Hill and R. I. M. Dunbar (2003), "Social Network Size in Humans," *Human Nature* 14(1): 53–72.

130. ***Locard's exchange principle:*** A. Rath, "Cyberpsychologist: Online, 'Every Contact Leaves a Trace,'" *All Things Considered,* National Public Radio, USA, npr.org, March 8, 2015.

132. **Nastiness online:** S. Rosenbloom, "Dealing with Digital Cruelty," *The New York Times,* August 23, 2014.

132. **Curt Schilling:** N. Golgowski and C. Red, "Former Red Sox Pitcher Curt Schilling Fires Back After Trolls' Violent, Sexual Tweets About Teen Daughter," *New York Daily News,* March 3, 2015.

133. **"made you feel uncomfortable":** S. Livingstone, L. Haddon, A. Görzig, and K. Ólafsson (2012), *EU Kids Online II: Final Report* (London: London School of Economics and Political Science), lse.ac.uk.

133. **Nearly ten thousand responses:** EU Kids Online (2014), *EU Kids Online: Findings. Methods. Recommendations* (London: London School of Economics and Political Science), lse.ac.uk.

137. **4.2 million pornography sites:** E. Mangan (n.d.), "How to Block Explicit Adult Content on Your Child's Internet and Mobile Devices—A Digital Parenting Guide," Digital Parenting.ie.

137. **2014 *Cosmopolitan* survey:** "This Is How You Watch Porn," Cosmopolitan .com, February 20, 2014.

138. **technology skills of their five-year-olds:** Ofcom, *Children and Parents: Media Use and Attitudes Report,* October 9, 2014, stakeholders.ofcom.org .uk.

139. **"hide their participation in risky":** "McAfee Digital Deception Study 2013: Exploring the Online Disconnect Between Parents & Pre-teens, Teens and Young Adults," mcafee.com, May 28, 2013.

142. **Terre des Hommes:** Terre des Hommes is a group of ten national organizations that works to further children's rights and equitable development: terredeshommes.org.

142. **"The moment [Sweetie] got online":** T. Ove, "Homeland Security Agents Tackle Growing Trend of Child Sex Tourism by Webcam," *Pittsburgh Post-Gazette,* January 1, 2014.

143. **"Cyberspace allows people":** J. R. Suler (2015), *Psychology of the Digital Age: Humans Become Electric* (Cambridge, U.K.: Cambridge University Press).

143. **When a sample of images:** NCMEC (2016), National Center for Missing and Exploited Children, missingkids.com.

143. **emergence of younger online offenders:** M. P. Aiken, M. Moran, and M. J. Berry (2011), "Child Abuse Material and the Internet: Cyberpsychology of Online Child-Related Sex Offending," www.interpol.int/Crime-areas/Crimes-against-children/Internet-crimes/.

144. **"early inappropriate sexualized event":** D. Howitt and K. Sheldon (2007), "The Role of Cognitive Distortions in Paedophilic Offending: Internet and Contact Offenders Compared," *Psychology, Crime & Law* 13(5): 469–86. See also L. Webb, J. Craissati, and S. Keen (2007), "Characteristics of Internet Child Pornography Offenders: A Comparison with Child Molesters," *Sex Abuse* 19: 449–65.

144. **the face of the man had been digitally swirled:** I. MacKinnon, "Thai Police Arrest Paedophile Suspect," *The Guardian,* October 19, 2007; "Alleged Pedophile Taught at B.C. School 6 Months Ago," cbc.ca/news, October 16, 2007.

147. **The first recorded death by robot:** D. Kravets, "Jan. 25, 1979: Robot Kills Human," *Wired,* January 25, 2010.

147. **advisory group in Ireland:** B. O'Neill, M. P. Aiken, M. Caffrey, J. Carthy, R. Lupton, A. Lynch, and K. O'Sullivan (2014), *Report of the Internet Content Governance Advisory Group,* Department of Communications, Energy

and Natural Resources, Ireland. See also T. Bouquet, "The Real-Life Spook Behind CSI: Cyber," telegraph.co.uk, November 3, 2015.

148. **they had seen pornographic images:** The One Poll survey for the NSPCC U.K. involved two thousand people between twelve and seventeen years of age, of whom seven hundred were aged twelve to thirteen. See NSPCC, "ChildLine Porn Campaign Confronts Issue of Young People and Porn," nspcc.org.uk, March 31, 2015.

149. **Content moderators of the Internet:** A. Chen, "The Laborers Who Keep Dick Pics and Beheadings out of Your Facebook Feed," Wired.com, October 23, 2014.

149. **"5 Deadly Sins":** M. P. Aiken, producer, "5 Deadly Sins," *CSI: Cyber,* CBS, J. Bruckheimer and P. Veasey, executive producers, aired March 6, 2016.

149. **hiding their online activity from their parents:** "McAfee Digital Deception Study 2013: Exploring the Online Disconnect Between Parents & Pre-teens, Teens and Young Adults," mcafee.com, May 28, 2013.

155. **Slender Man:** A. Jones, "The Girls Who Tried to Kill for Slender Man," *Newsweek,* August 13, 2014.

155. **"This should be a wake-up call":** R. P. Jack, "Waukesha Stabbing News Conference," *Los Angeles Times,* June 2, 2014.

157. **Resilence:** (n.d.) "The Road to Resilience," American Psychological Association, apa.org.

158. **"a growing unease that the goal of risk prevention":** E. Staksrud and S. Livingstone (2009), "Children and Online Risk: Powerless Victims or Resourceful Participants?," *Information, Communication and Society* 12(3): 364–87.

161. **I raised this in an editorial:** M. Aiken, "Parents Alone Cannot Police Our Youth in Cyberspace," Irish *Independent,* September 23, 2014.

161. ***Child abuse* is defined:** World Health Organization (1999), "Report of the Consultation on Child Abuse Prevention" (WHO, Geneva, March 29–31, 1999), 15.

162. **The four main principles:** United Nations, *Convention on the Rights of the Child* (New York, November 20, 1989).

163. **I formally proposed that a new amendment be considered:** M. P. Aiken (2015), "Children's Rights in Cyberspace," The Hague Talks: Setting Peace and Justice in Motion. We, the People: The United Nations at 70—The Global Impact of the Digital Age (October 23), The Hague, The Netherlands.

 See also "Universal Children's Day announcement—Consultation period regarding proposed amendment to the 1989 UN Convention on the Rights of the Child to incorporate the Cyber Rights of the Child," November 20, 2015, haguejusticeportal.net.

Chapter 5: Teenagers, Monkeys, and Mirrors

165. **Inside her classroom:** L. O'Neil, "Student's Selfie with Pregnant Teacher 'in Labour' Goes Viral," cbc.ca, October 18, 2013.
168. **"SELFIE-ISH":** "SELFIE-ISH! My Photo with Brooklyn Bridge Suicide Dude," *New York Post* (2013), reproduced in J. Jones, "Don't Hate the Woman Behind the 'World's Worst Selfie,' " The Guardian.com, December 5, 2013.
169. **Narcissistic Personality Inventory:** R. N. Raskin and C. S. Hall (1979), "A Narcissistic Personality Inventory," *Psychological Reports* 45(2): 590; and R. Raskin and H. Terry (1988), "A Principal-Components Analysis of the Narcissistic Personality Inventory and Further Evidence of Its Construct Validity," *Journal of Personality and Social Psychology* 54(5): 890–902.
170. ***mirror-image stimulation:*** G. G. Gallup (1977), "Self-Recognition in Primates: A Comparative Approach to the Bidirectional Properties of Consciousness," *American Psychologist* 32(5): 329–38, quotation from 330.
170. ***self-concept:*** S. A. McLeod (2008), "Self Concept," *Simply Psychology,* quoting from R. F. Baumeister (ed.) (1999), *The Self in Social Psychology.* simplypsychology.org.

170–71. **The isolates remained fascinated:** G. G. Gallup and M. K. McClure (1971), "Preference for Mirror-Image Stimulation in Differentially Reared Rhesus Monkeys," *Journal of Comparative and Physiological Psychology* 75(3): 403–7.

171. **self-recognition:** B. Amsterdam (1972), "Mirror Self-Image Reactions Before Age Two," *Developmental Psychobiology* 5(4): 297–305; A. H. Schulman and C. Kaplowitz (1977), "Mirror-Image Response During the First Two Years of Life," *Developmental Psychobiology* 10(3):133–42.
171. **He described self-concept:** C. R. Rogers (1959), "A Theory of Therapy, Personality, and Interpersonal Relationships, As Developed in the Client-Centered Framework," in S. Koch (ed.), *Psychology: A Study of a Science. Study 1, Volume 3: Formulations of the Person and the Social Context* (New York: McGraw-Hill), 184–256.
172. **"the cyber self":** In the earliest days of digital presence, psychologists Hazel Markus and Paula Nurius presciently saw the need down the road for a conversation about the virtual self, given the beginnings of avatars and talk of migrating our identities and lives online. In an article written for *The Atlantic* in 1987, they described the appearance in the virtual world of the "Possible Self."
173. **as many "selves" as we have friends:** W. James (1892), *Psychology: Briefer Course* (New York: Henry Holt and Company), especially 179–80.
173. **Presenting yourself to the world:** T. J. Scheff (1988), "Shame and Confor-

mity: The Deference-Emotion System," *American Sociological Review* 53(3): 395–406, especially 396.

174. **Narcissus:** The earliest surviving versions of the story of Narcissus were thought to be related by Konon in Greek (*Narratives*, 24) and Ovid in Latin (*Metamorphoses,* Book III, 342–510). However, the unlikely survival of a papyrus from an Egyptian rubbish dump shows that the story was in circulation at least half a century earlier, in the mid–first century B.C.: D. Keys (2004), "The Ugly End of Narcissus," *BBC History Magazine* 5(5): 9.

175. ***identity versus role confusion***: S. McLeod (2013), "Erik Erikson," simply psychology.org.

175. **a significant increase in scores:** J. M. Twenge and J. D. Foster (2010), "Birth Cohort Increases in Narcissistic Personality Traits Among American College Students, 1982–2009," *Social Psychological and Personality Science* 1(1): 99–106.

175–76. **parents who have overpraised their children:** E. Brummelman et al. (2015), "Origins of Narcissism in Children," *Proceedings of the National Academy of Sciences* 112(12): 3659–62.

176. ***The Picture of Dorian Gray:*** The story was first published in thirteen chapters in *Lippincott's* magazine in 1890. That version was rejected by Britain's biggest bookseller because of its sexual overtones, so a more muted, twenty-chapter version of the novel was published in 1891; even this was considered salacious enough to count against Wilde when he was put on trial for gross indecency in 1895. Incredibly, it was not until 2011 that the unexpurgated text was published. See the account of the book's early history on the British Library website.

177. **Now there's a plethora of apps:** P. Samotin, "The 7 Best Photo Editing Apps to Look Better in Every Selfie," stylecaster.com, March 13, 2014.

179. **"Youth is subjected by our civilization . . .":** J. Zimmerman, "Schools Can't Stop Kids from Sexting. More Technology Can," nytimes.com, November 10, 2015.

179. **These images are hacked and stolen:** C. Cullen, " 'Unless You Never Upload a Photo to Social Media Ever, You're Not Safe': Irish Teens Speak of Photo Hacking Horror," Irish *Independent,* January 14, 2016.

180. **rise in plastic surgery:** American Academy of Facial Plastic and Reconstructive Surgery: "Selfie Trend Increases Demand for Facial Plastic Surgery," aafprs.org, March 11, 2014.

181. **Online, social capital can be important:** C. Steinfield, N. B. Ellison, C. Lampe, and J. Vitak (2013), "Online Social Network Sites and the Concept of Social Capital," in F. L. F. Lee et al. (eds.), *Frontiers in New Media Research* (New York: Routledge), 115–31; and N. Lin (1999), "Building a Network Theory of Social Capital," *Connections* 22(1): 28–51.

182. **"I broke down in tears":** T. Willis and A. Bevan, "Tallulah Willis Opens Up: Her Struggle with Substance Abuse and Self-Esteem," *TeenVogue,* January 7, 2015.

183. **boys and young men suffering from eating disorders:** D. Cox, "Are More Men Getting Eating Disorders?," *The Guardian,* January 18, 2015.

184. **"I was constantly in search":** G. Aldridge and K. Harden, "Selfie Addict Took TWO HUNDRED a Day—and Tried to Kill Himself When He Couldn't Take Perfect Photo," mirror.co.uk, March 23, 2014.

185. **body dysmorphic disorder:** M. Tartakovsky (2013), "Body Dysmorphic Disorder: When the Reflection Is Revolting," Psych Central, January 30, 2013.

185. **a list of some common signs:** Mayo Clinic Staff (n.d.), "Body Dysmorphic Disorder: Symptoms and Causes," mayoclinic.org

187. **"self-actualization":** C. R. Rogers (1951), *Client-Centered Therapy* (Boston: Houghton Mifflin Harcourt).

187. **"I find it easier to be myself on the Internet":** S. Livingstone, L. Haddon, A. Görzig, and K. Ólafsson (2011), "Risks and Safety on the Internet: The Perspective of European Children. Full Findings and Policy Implications from the EU Kids Online Survey of 9–16 Year Olds and Their Parents in 25 Countries," (London: LSE), 40.

188. **"The digital self becomes less":** C. Moss, "Men and Boys Are in Crisis, and Technology Is to Blame," *The Telegraph,* May 10, 2015.

189. **"There was absolutely nothing wrong":** "Teens Threatened with 'Sexting' Porn Charge Sue Prosecutor," *The Independent,* March 26, 2009.

189. **legal implications:** J. L. Barry (2010), "The Child as Victim and Perpetrator: Laws Punishing Juvenile 'Sexting,'" *Vanderbilt Journal of Entertainment and Technology Law* 13(1): 129–53.

189. **"the production and dissemination":** M. G. Leary (2010), "Sexting or Self-Produced Child Pornography—The Dialogue Continues . . . ," *Virginia Journal of Social Policy and the Law* 17(3), 488.

189. **In Australia, child pornography laws:** J. Tin, "More Than 450 Child Pornography Charges Laid Against Youths Aged 10 to 17 in Past Three Years," *The Sunday Mail* (Qld), October 9, 2011.

189. **incidence of sexting:** The National Campaign to Prevent Teen and Unplanned Pregnancy (2008), *Sex and Tech: Results from a Survey of Teens and Young Adults* (Washington, DC).

189. **half of all teenagers having sent a sext:** H. Strohmaier, M. Murphy, and D. DeMatteo (2014), "Youth Sexting: Prevalence Rates, Driving Motivations, and the Deterrent Effect of Legal Consequences," *Sexuality Research and Social Policy* 11(3): 245–55. See also the report on the Drexel study and surrounding issues by K. Wallace, "Chances Are, Your Teen Has Sexted," edition.cnn.com, January 2, 2015.

190. **In their seminal article:** D. Finkelhor and J. Wolak, "Sext and Sensibility," huffingtonpost.com, July 11, 2012.

192. **Jesse Logan:** M. Celizic, "Her Teen Committed Suicide Over 'Sexting,'" today.com, March 6, 2009.

193. **sextortion:** "Amanda Todd Suicide: RCMP Repeatedly Told of Blackmailer's Attempts," cbc.ca/news, November 15, 2013.

194. **a thirty-five-year-old man:** "Man Charged in Netherlands in Amanda Todd Suicide Case," BBC.com, April 18, 2014.

196. **default privacy settings:** R. Gross and A. Acquisti, "Information Revelation and Privacy in Online Social Networks," in Association for Computing Machinery, *WPES '05: Proceedings of the 2005 ACM Workshop on Privacy in the Electronic Society* (Alexandria, VA, November 7, 2005), 71–80.

196. **"privacy" concerns:** For more about online sharing, see E. Segran, "The Truth About Teenagers, the Internet, and Privacy," fastcompany.com, November 4, 2014.

197. ***Groupthink:*** I. L. Janis (1971), "Groupthink," *Psychology Today* 5(3); reprinted in H. J. Leavitt, L. R. Pondy, and D. M. Boje (Eds) (1980), *Readings in Managerial Psychology* (3rd edition, Chicago: University of Chicago Press), 432–44.

198. **"There isn't a school in the United States":** M. Martinez, "Sexting Scandal: Colorado High School Faces Felony Investigation," edition.cnn.com, November 9, 2015.

199. **"She was a 16-year-old California girl":** K. Poulsen, "Pimps Go Online to Lure Kids into Prostitution," *Wired,* February 25, 2009.

200. **"He showed me the ropes":** Y. Alcindor, "Sex Trafficking in the USA Hits Close to Home," usatoday30.usatoday.com, September 27, 2012.

200. **technology has created a new dimension:** M. P. Aiken and S. Chan (2015), "Cyber Criminology: Algorithmic *vs.* Heuristical Approaches for Analysis Within the Human Trafficking Domain," *International Journal of Advancements in Technology* 6(2); and S. Chan and M. P. Aiken, "Big Data to Big Insight for Human Trafficking: Analysis of Adaption Cycles for Digital Image Deployment on Online Classified Sites," paper presented at White House Forum to Combat Human Trafficking (West Wing, Washington, DC), April 9, 2013.

201. **online classified adult advertising:** "Tulsa Police: Online Escort Ads a Front for Prostitution," washingtontimes.com, August 17, 2015.

201. **Backpage picked up where Craigslist left off:** B. Thompson, "Many Believe Backpage.com Is Online Portal for Human Trafficking," miheadlines.com, March 16, 2014.

202. **proceedings against Backpage:** E. Koh, "Dallas-Based Backpage.com Faces

Civil Contempt Charges in Senate Sex-Trafficking Inquiry," *The Dallas Morning News,* February 11, 2016.

202. **While they supposedly can disappear:** V. Woollaston, "Gone but NOT Forgotten: 'Deleted' Snapchat Photos Are Stored on Your Phone and Can Be Easily Downloaded, Forensics Firm Claims," dailymail.co.uk, May 10, 2013.

Chapter 6: Cyber Romance

208. **the "science of love":** H. Song et al. (2015), "Love-Related Changes in the Brain: A Resting-State Functional Magnetic Resonance Imaging Study," *Frontiers in Human Neuroscience,* February 13.
211. **increase in online-dating-related rape:** *Emerging new threat in online dating: Initial trends in internet dating-initiated serious sexual assaults,* national crimeagency.gov.uk, February 7, 2016.
211. **those who met on the Internet:** J. T. Cacioppo et al. (2013), "Marital Satisfaction and Break-ups Differ Across On-line and Off-line Meeting Venues," *Proceedings of the National Academy of Sciences* 110(25): 10135–40.
212. **online dating . . . $2.2 billion by 2015:** "Dating Services in the US: Market Research Report," ibisworld.com, April 2016.
212. **heightened in the future:** S. Curtis, "DNA Matching and Virtual Reality: The World of Online Dating in 2040," *The Telegraph,* November 27, 2015.
212. **dating app Tinder:** gotinder.com (2016).
214. **"hidden desires and fears":** R. U. Akeret (2000), *Photolanguage: How Photos Reveal the Fascinating Stories of Our Lives and Relationships* (London: W. W. Norton & Company).
215. **Making a quick judgment:** N. Ellison, R. Heino, and J. Gibbs (2006), "Managing Impressions Online: Self-Presentation Processes in the Online Dating Environment," *Journal of Computer-Mediated Communication* 11(2).
215. **This is evolutionary reality:** P. M. Todd, L. Penke, B. Fasolo, and A. P. Lenton (2007), "Different Cognitive Processes Underlie Human Mate Choices and Mate Preferences," *Proceedings of the National Academy of Sciences of the United States of America* 104 (38): 15011–16. See also N. Barber (1995), "The Evolutionary Psychology of Physical Attractiveness: Sexual Selection and Human Morphology," *Ethology and Sociobiology* 16(5): 395–424.
215. **"desirability score":** C. Macdonald, "What's YOUR Secret Tinder Score?: CEO Reveals App Has an 'Internal Rating' Used to Select Matches," daily mail.co.uk, January 11, 2016.

216. ***impression management:*** "Impression management is an active self presentation of a person aiming to enhance his image in the eyes of others": J. B. P. Sinha (2008), *Culture and Organizational Behaviour* (New Delhi: Sage Publications), 104. The theory started with Erving Goffman (1959), *The Presentation of Self in Everyday Life* (New York: Anchor). He used the metaphor of a theatrical performance to explain how we present ourselves and attempt to guide the impressions others form about us.

216. **what kind of profile photographs:** K. Lee, "The Research and Science Behind Finding Your Best Profile Picture," blog.bufferapp.com, March 25, 2015.

217. **"Squinch":** P. Hurley (2013), *"It's All About the Squinch!"* YouTube.com.

217. **A famous case of . . . *identity deception*:** The Alex/Joan story was originally reported by L. Van Gelder, "The Strange Case of the Electronic Lover," *Ms.* magazine, October 1985. For more about online communication and impression management, see A. Chester and D. Bretherton (2007), "Impression Management and Identity Online," in A. Joinson et al. (eds.), *The Oxford Handbook of Internet Psychology* (Oxford: Oxford University Press), 223–36. On deception, see M. Whitty and A. Joinson (2009), *Truth, Lies and Trust on the Internet* (New York: Routledge).

218. ***catfishing:*** *Catfish* (2011), the original documentary, was directed by Ariel Schulman and Henry Joost. The story's protagonist later wrote a book about his experiences with advice for the millions of viewers of his MTV show, *Catfish,* on approaching digital relationships: N. Schulman (2014), *In Real Life: Love, Lies & Identity in the Digital Age* (New York: Hachette). On the evolution of the term *catfishing,* see A. Harris, "Who Coined the Term 'Catfish'?," Slate.com, January 18, 2013.

218. ***trolls:*** Research identifies the characteristics of trolling as "aggression, deception, disruption, and success." See C. Hardaker (2010), "Trolling in Asynchronous Computer-Mediated Communication: From User Discussions to Academic Definitions," *Journal of Politeness Research* 6: 215–42.

219. **A smart solution to catfishing:** N. Lomas, "Dating App Newbie Blume Wants to Kill Catfishing with Ephemeral Selfies," techcrunch.com, January 16, 2016.

219. **"We're definitely not trying to be sexist":** A. Shontell, "What It's Like to Found a $750 Million Startup, Go Through a Sexual-Harassment Lawsuit, and Start All Over by Age 25," uk.businessinsider.com, January 27, 2015.

220. ***dark tetrad of personality:*** E. E. Buckels, P. D. Trapnell, and D. L. Paulhus (2014), "Trolls Just Want to Have Fun," *Personality and Individual Differences* 67: 97–102.

220. **"narcissistic personality disorder":** mayoclinic.org, November 18, 2014.

222. **"assign their first born child":** The June 2014 experiment was carried out for the Finnish ethical computer security company F-Secure. The company notes drily that "while terms and conditions are legally binding—it is contrary to public policy to sell children in return for free services, so the clause would not be enforceable in a court of law." See F-Secure (2014), *Tainted Love: How Wi-Fi Betrays Us,* f-secure.com.

222. **It is well known in military psychology:** M. Coeckelbergh, "War from a Distance: The Ethics of Killer Robots," *E-International Relations,* June 16, 2014. Coeckelbergh points out that distance blunts our sense of responsibility. At a distance, "the natural moral-psychological barrier to killing is removed. There is no place for empathy, no knowledge of the suffering one causes, and the distance between killer and target seems unbridgeable." For an in-depth study of this phenomenon, see D. Grossman (1996), *The Psychological Cost of Learning to Kill in War and Society* (New York: Back Bay Books), section III.

223. **The hacking of Ashley Madison:** A. Krasodomski-Jones, "Role Play and Bubble Baths: Analysing the Ashley Madison Hack," wired.co.uk, September 7, 2015.

224. **a Facebook affair or cyber-infidelity:** J. Insley, "Cyber Affairs Cited in Breakdown of Real Marriages," theguardian.com, May 27, 2009.

224. **cyber-infidelity is a burgeoning area of the law:** A. Feldstein, "Is Cybersex Grounds for Divorce?," huffingtonpost.com, March 10, 2014.

224. **outcomes of virtual distractions and disloyalties:** J. P. Schneider, R. Weiss, and C. Samenow (2012), "Is It Really Cheating?: Understanding the Emotional Reactions and Clinical Treatment of Spouses and Partners Affected by Cybersex Infidelity," *Sexual Addiction and Compulsivity. The Journal of Treatment and Prevention: Cyber Sex* 19(1–2): 123–39.

225. **cyber-jealousy of Giuseppe Castro:** S. Nelson, "Jealous husband 'decapitated his wife over online affair,'" dailymail.co.uk, April 3, 2009.

225. ***LovePlus* and its sequels:** B. Bosker, "Meet the World's Most Loving Girlfriends—Who Also Happen to Be Video Games," huffingtonpost.com, January 25, 2014.

225. **people can form real and authentic emotional attachments to virtual characters:** M. Coulson, J. Barnett, C. J. Ferguson, and R. L. Gould (2012), "Real Feelings for Virtual People: Emotional Attachments and Interpersonal Attraction in Video Games," *Psychology of Popular Media Culture* 1(3): 176–84.

226. **nearly 40 percent of Japanese men and women:** Y. Wakatsuki (2015), "Middle-aged Virgins: Why So Many Japanese Stay Chaste," edition.cnn.com, June 24, 2015.

227. **"These days, insecure in our relationships":** S. Turkle (2011), *Alone To-*

gether: Why We Expect More from Technology and Less from Each Other (New York: Basic Books).

227. **"Strangely enough, our brains don't seem to care . . .":** C. Malkin, "How Technology Makes Us Afraid of Intimacy," huffingtonpost.com, November 23, 2012.
228. **but less actual contact sex:** T. Wayne, "With Some Dating Apps: Less Casual Sex Than Casual Text," *New York Times,* November 7, 2014.
228. **The field of cyber-love and robotics:** G. Gurley, "Is This the Dawn of the Sexbots? (NSFW)," *Vanity Fair,* April 2015.
228. **unusual step of creating an urgent "term of condition":** R. Patel, "Do Not Have Sex with Our Robots, Japanese Firm Warns Users," *International Business Times,* September 28, 2015.
229. **"Pixar took a lesson from *Tin Toy*":** J. Hsu, "Why 'Uncanny Valley' Human Look-Alikes Put Us on Edge," *Scientific American,* April 3, 2012.
229. ***Uncanny Valley:*** M. Mori (1970), "The Uncanny Valley," *Energy* 7(4): 33–35; trans. K. F. MacDorman and N. Kageki, *Spectrum,* June 12, 2012. For Mori's reflections on the "Uncanny Valley" phenomenon, see recent interview with him by N. Kageki, "An Uncanny Mind: Masahiro Mori on the Uncanny Valley and Beyond," *Spectrum,* June 12, 2012, spectrum.ieee.org.
230. **he first experienced what he described as an "eerie sensation":** M. Mori describes how he became aware of the Uncanny Valley in *Uncanny Valley Revisited: Masahiro Mori,* International Conference on Intelligent Robots and Systems, Tokyo International Exhibition Centre, Japan, November 6, 2013: YouTube.com, November 20, 2013.
230. **Even monkeys appear to have an Uncanny Valley reflex:** S. A. Steckenfinger and A. A. Ghazanfar (2009), "Monkey Visual Behavior Falls into the Uncanny Valley," *Proceedings of the National Academy of Sciences* 106(43): 18362–66, pnas.org.

Chapter 7: Cyberchondria and the Worried Well

232. **Clutching their "Google stack":** M. Aiken and G. Kirwan (2012), "Prognoses for Diagnoses: Medical Search Online and 'Cyberchondria,'" *BMC Proceedings* 6 (suppl. 4): 30.
233. **unnecessary medical visits:** A. Nazaryan, "Internal Affairs: On Hypochondria," *The New Yorker,* August 2, 2012.
235. **In a large international survey:** D. McDaid and A. Park (2011), *BUPA Health Pulse 2010. Online Health: Untangling the Web,* 3: bupa.com.au. See also *Bupa Health Pulse 2011: International Healthcare Survey. Global Trends, Attitudes and Influences,* 16: bupa.com.au.
235. **"Health and long life to you":** The Old Irish blessing has many permuta-

tions; this version probably derives from the time of the Great Famine (1845–1852), when land confiscation, high rents, and the potato blight forced mass Irish emigration. The phrase "and death in Old Ireland" expresses the wish that immigrants might live to return to their home country.

235. **When does a normal desire for health:** M. P. Aiken and G. H. Kirwan (2014), "The Psychology of Cyberchondria and 'Cyberchondria by Proxy,' " in A. Power and G. Kirwan (eds.), *Cyberpsychology and New Media: A Thematic Reader* (New York: Psychology Press), 158–69. See also C. McMahon and M. P. Aiken (2015), "Introducing Digital Wellness: Bringing Cyberpsychological Balance to Healthcare and Information Technology," in *Proceedings of the 14th IEEE International Conference on Ubiquitous Computing and Communications,* computer.org/cps, 1417–22.

236. **people make judgments online in a number of ways:** C. L. Corritore, B. Kracher, and S. Wiedenbeck (2003), "On-line Trust: Concepts, Evolving Themes, a Model," *International Journal of Human-Computer Studies* 58: 737–58. See also L. C. Vega, T. DeHart, and E. Montague (2011), "Trust Between Patients and Health Websites: A Review of the Literature and Derived Outcomes from Empirical Studies," *Health and Technology,* 1(2–4): 71–80.

236. **the criteria that online searchers use:** A. N. Joinson, K. Y. A. McKenna, T. Postmes, and U-D. Reips (eds.) (2007), *The Oxford Handbook of Internet Psychology* (Oxford: Oxford University Press), 351.

236. **online medical websites:** WebMD.com, NIH.gov, and MayoClinic.org monthly unique visitor metrics: alexa.com (2016).

237. **people killed by lightning strikes:** There were 235 lightning fatalities in the United States in the years 1999–2003, an average of 47 people per year, according to the National Oceanic and Atmospheric Administration's Weather Service. Interestingly, the average has dropped to 26 per year in the period 2011–2015, per nws.noaa.gov.

238. **"an active center for Online Support Groups":** mdjunction.com.

239. **Lyme might be passed from a pregnant mother:** M. Lavelle, "Mothers May Pass Lyme Disease to Children in the Womb . . . but Public Health Experts Say the Science Isn't So Clear," ScientificAmerican.com, September 22, 2014.

239. **the disease could be sexually transmitted:** "Recent Study Suggests That Lyme Disease Can Be Sexually Transmitted," lymedisease.org, January 25, 2014. See also Centres for Disease Control and Prevention, "Lyme Disease Frequently Asked Questions: Can Lyme Disease be Transmitted Sexually?," cdc.gov.

240. **Medical misinformation can also leak:** S. Usborne, "Cyberchondria: The Perils of Internet Self-Diagnosis," *The Independent,* February 17, 2009.

240. **A Pew Internet Project study:** S. Fox and M. Duggan, "Health Online 2013," *Pew Research Internet Project,* 3. pewinternet.org.
242. **The term *cyberchondria*:** " 'Cyberchondria' Hits Web Users," BBC World News, April 13, 2011. The Harris Poll first used the word *cyberchondriac* to describe people in 1998, when just over 50 million American adults had gone online to look for health information: #11, February 17, 1999. By 2005, that number had risen to 117 million. In the 2010 poll, the number of cyberchondriacs had jumped to 175 million from 154 million the previous year: " 'Cyberchondriacs' on the Rise?," theharrispoll.com, August 4, 2010. See also J. Stone and M. Sharp (2003), "Internet Resources for Psychiatry and Neuropsychiatry," *Journal of Neurology, Neurosurgery and Psychiatry* 74(1): 10–12.
242. **a groundbreaking study:** R. W. White and E. Horvitz (2009), "Cyberchondria: Studies of the Escalation of Medical Concerns in Web Search," *ACM Transactions on Information Systems* 27(4), Article 23.
242. **"health anxiety":** G. J. G. Asmundson, S. Taylor, S. Sevgur, and B. J. Cox (2001), "Health Anxiety: Classification and Clinical Features," in G. J. G. Asmundson, S. Taylor, and B. J. Cox (eds.) (2001), *Health Anxiety: Clinical and Research Perspectives on Hypochondriasis and Related Conditions* (Chichester, U.K.: John Wiley & Sons), 4–5.
243. **The desire for information that will help us thrive is a natural, primal urge:** J. Panksepp (2003), "An Archaeology of Mind: The Ancestral Sources of Human Feelings," *Soundings: An Interdisciplinary Journal* 86(1/2): 41–69.
245. **a number of sweeping changes:** American Psychiatric Association (2013), *Diagnostic and Statistical Manual of Mental Disorders,* 5th ed. (Arlington, VA: American Psychiatric Association), 812–13.
246. **anxiety and concern:** C. F. Belling (2006), "Hypochondriac Hermeneutics: Medicine and the Anxiety of Interpretation," *Literature and Medicine* 25(2): 376–401.
246. **Hypochondria is found in 4 to 9 percent:** The prevalence of somatic symptom disorder is not known. The DSM-5 estimates the figure to be around 5–7 percent of the adult population, while it puts prevalence of illness anxiety disorder between 1.3 and 10 percent (pp. 312, 316). Although there is no comprehensive international study, some research has been conducted in individual countries, for example, Australia, which is at 5.7 percent (M. Sunderland, J. M. Newby, and G. Andrews [2013], "Health Anxiety in Australia: Prevalence, Comorbidity, Disability, and Service Use," *The British Journal of Psychiatry* 202(1): 56–61, and Germany, at around 6 percent (G. Bleichhardt and W. Hiller [2007], "Hypochondriasis and Health Anxiety in the German Population," *British Journal of Health Psychology* 12[4]: 511–23).
247. **a wide range of emotional difficulties:** T. A. Widiger and P. T. J. Costa

(1994), "Personality and Personality Disorders," *Journal of Abnormal Psychology* 103(1): 78–91.

247. **average length of a consultation:** E. Sillence, P. Briggs, P. Harris, and L. Fishwick (2006), "Changes in Online Health Usage over the Last 5 Years," in *CHI '06 Extended Abstracts on Human Factors in Computing Systems* (New York: ACM Press), 1331–36.

248. **suffering from "Munchausen syndrome":** R. Asher (1951), "Munchausen's Syndrome," *Lancet* 1(6650): 339–41.

248. **"Munchausen syndrome by proxy":** R. Meadow (1977), "Munchausen Syndrome by Proxy: The Hinterland of Child Abuse," *Lancet* 2(8033): 343–45.

248. **proliferation of various conditions:** L. J. Lasher and M. S. Sheridan (2013), *Munchausen by Proxy: Identification, Intervention, and Case Management* (New York: Routledge).

248. **also known as "hospital addicts":** D. O. Day and R. L. Moseley (2010), "Munchausen by Proxy Syndrome," *Journal of Forensic Psychology Practice* 10(1): 13–36.

248–49. **develops the condition after hospitalization for a true illness:** L. Criddle (2010), "Monsters in the Closet: Munchausen Syndrome by Proxy," *Critical Care Nurse* 30(6): 46–55.

250. **evolution of the original condition:** J. Kleeman, "Sick Note: Faking Illness Online," *The Guardian,* February 26, 2011.

251. **he found exactly what he was looking for online:** A. Moses, "Alarm Sounded over Dr Google's Diagnosis," *The Sydney Morning Herald,* February 10, 2011.

252. **more dependent on online searches:** Y. Amichai-Hamburger (2007), "Personality, Individual Differences and Internet Use," in A. M. Joinson, K. Y. A. McKenna, T. Postmes, and U-D. Reips (eds.), *The Oxford Handbook of Internet Psychology* (Oxford: Oxford University Press), 187–204.

252. **In 2013, more than 1,600:** D. Fine Maron, "Pill of Goods: International Counterfeit Drug Ring Hit in Massive Sting," ScientificAmerican.com, July 3, 2013.

255. **missing common cancers:** G. Kolata (2005), "Rapid Rise and Fall for Body-Scanning Clinics," *The New York Times,* January 23, 2005.

255. **"You don't know what's inside until you look":** W. James, "Know Thy Inner Self: A 3-D Scan Spots Cancer and Other Health Risks Before It's Too Late," bodyscanintl.com.

256. **He was celebrated on *The Oprah Winfrey Show*:** M. Ballon, "Full-Body Scan Pioneer Is on the Outside Looking In," *Los Angeles Times,* August 19, 2001.

256. **evolving pathology was found in *every case:*** P. Bowes, "US Doctors Offer Full Body Scan," news.bbc.co.uk, January 2, 2001.

256. **"It's a daily event":** R. Davis, "The Inside Story," *USA Today*, August 25, 2000.

256. **"Yes, there's an increased risk":** P. Eastman (2005), "Whole-Body Scanning for Patients with No Symptoms: What Are the Pros and Cons?," *Oncology Times UK* 2(3): 20–21.

256. **third most common cause of death:** B. Starfield (2000), "Is US Health Really the Best in the World?," *The Journal of the American Medical Association* 284(4): 483–85.

257. **deaths . . . a result of medical error in hospitals:** L. T. Kohn, J. M. Corrigan, and M. S. Donaldson (eds.) (2000), *To Err Is Human: Building a Safer Health System* (Washington, DC: National Academy Press); nap.edu.

257. **180,000 patient deaths annually:** D. R. Levinson (2010), *Adverse Events in Hospitals: National Incidence Among Medicare Beneficiaries* (Office of Inspector General, U.S. Department of Health and Human Services).

257. **stunning jump in deaths:** J. T. James (2013), "A New, Evidence-based Estimate of Patient Harms Associated with Hospital Care," *Journal of Patient Safety* 9(3): 122–28. The report used a weighted average of four studies carried out between 2008 and 2011 to reach the lower limit of 210,000; the higher figure of 440,000 was arrived at by estimating errors and omissions in medical records and diagnostic procedures (p. 127). Now see M. A. Makary and M. Daniel (2016), "Medical Error—the third leading cause of death in the US," *British Medical Journal* 353.

258. **when as many as 60 percent of Americans search for health information:** S. Fox and M. Duggan, "Health Online 2013," PewInternet.org, January 15, 2013.

259. **The Google motto "Don't be evil":** Google's Code of Conduct is prefaced with: "Don't be evil. We believe strongly that in the long term, we will be better served—as shareholders and in all other ways—by a company that does good things for the world even if we forgo some short term gains," Securities and Exchange Commission registration statement, sec.gov/Archives.

259. **"First, do no harm":** This phrase does not actually occur in the classical Hippocratic Oath. The closest phrase is: "I will follow that system of regimen which, according to my ability and judgment, I consider for the benefit of my patients, and abstain from whatever is deleterious and mischievous." But modern versions of the oath often have something more general. The Latin phrase is attributed to the seventeenth-century English physician Thomas Sydenham, in a book by surgeon and mythologist Thomas Inman from 1860 called *Foundation for a New Theory and Practice of Medicine* (London: John Churchill). On page 244 it states: "We crouch under the cloak of Sydenham, and say, that our motto is none other than a translation of his Latin aphorism respecting a physician's duties, viz:—'*Primum est ut non nocere.*'"

259. **This finds agreement in Japan:** In the past the decision was left up to the patient's family, who were usually reluctant to disclose the truth, but in the 1980s the introduction of hospice care and new, aggressive treatments resulted in a government recommendation that the terminally ill patient be given both a diagnosis and an estimate of life expectancy. The National Cancer Center issued "Guidelines for Telling the Truth to Cancer Patients" (H. Okamura et al. [1998], *Japanese Journal of Clinical Oncology* 28[1]: 1–4). Included were instructions that may seem basic—such as "On no account should the diagnosis be communicated via the telephone, or in passing in a corridor, or in a public place"—but that reveal how unaccustomed Japanese physicians had been to approaching such a task (Y. Uchitomi and S. Yamawaki [1997], "Truth-telling Practice in Cancer Care in Japan," *Annals of the New York Academy of Sciences* 809: 290–99).

Chapter 8: What Lies Beneath: The Deep Web

261. **Formal or informal codes of conduct:** Group socialization is described in R. Moreland and J. Levine (1982), "Socialization in Small Groups: Temporal Changes in Individual-Group Relations," in L. Berkowitz (ed.), *Advances in Experimental Social Psychology* vol. 15 (New York: Academic Press), 137–92.

262. **the golden age of piracy:** The privateers and buccaneers who took their orders from governments seeking trade routes and scores against the vessels of rival nations lost their commissions in peacetime; many put their skills to use as pirates, though their careers were often limited to one or two years. See thewayofthepirates.com (2016).

262. **lawlessness that impacts all of us:** M. P. Aiken, C. McMahon, C. Haughton, L. O'Neill, and E. O'Carroll (2015), "A Consideration of the Social Impact of Cybercrime: Examples from Hacking, Piracy, and Child Abuse Material Online," *Contemporary Social Science: Journal of the Academy of Social Sciences*. DOI: 10. 1080/21582041.2015.1117648.

264. **"free" virus-laden malware:** "Social engineering is a component of many—if not most—types of exploits. Virus writers use social engineering tactics to persuade people to run malware-laden email attachments, phishers use social engineering to convince people to divulge sensitive information, and scareware vendors use social engineering to frighten people into running software that is useless at best and dangerous at worst": S. Guvakuva, "What is Social Engineering," technomag.co.zw, December 3, 2015.

264. **the Deep Web refers to:** Dr. Jill Ellsworth first used the term *invisible Web* to refer to poorly marketed websites such as one reasonably designed but not registered with any of the search engines, so no one can find it. See F. Garcia (1996), "Business and Marketing on the Internet," *TCP Online*

9(1), January (accessed August 24, 2015, via the Wayback Machine Internet Archive). Difficulties with this term and the first use of the now-preferred *Deep Web* are to be found in the study of Internet expert Mark Bergman (2001), "White Paper: The Deep Web: Surfacing Hidden Value," *The Journal of Electronic Publishing* 7(1).

266. **Darknets refer to what is deliberately hidden:** A. Greenberg, "Hacker Lexicon: What Is the Dark Web?," *Wired,* November 19, 2014.

266. **digital black market has grown:** "Going Dark: The Internet Behind the Internet," National Public Radio, May 25, 2014, npr.org.

266. **one Darknet tutorial:** "How to Access THE HIDDEN WIKI (Deep Web, Secret Internet)," TheBot.net, March 20, 2011.

269. **McDumpals, one of the leading sites:** B. Krebs, "Peek Inside a Professional Carding Shop," krebsonsecurity.com, June 14, 2014.

269. **gamble on the actual time of execution:** "The Disturbing World of the Deep Web, Where Contract Killers and Drug Dealers Ply Their Trade on the Internet," dailymail.co.uk, October 11, 2013.

269. **Ross Ulbricht:** The biography of Ross Ulbricht and the story of the Silk Road is taken from D. Kushner, "Dead End on Silk Road: Internet Crime Kingpin Ross Ulbricht's Big Fall," *Rolling Stone,* February 4, 2014; and M. J. Barratt, J. A. Ferris, and A. R. Winstock (2014), "Use of Silk Road, the Online Drug Marketplace, in the United Kingdom, Australia and the United States," *Addiction* 109: 774–83.

270. **the writings of Austrian economist Ludwig von Mises:** For the life and work of economist and social philosopher Ludwig von Mises, see the profile at the Institute bearing his name: mises.org/profile/ludwig-von-mises; and A. Carden, "The Greatest Thinker You've Never Read: Ludwig von Mises," *Forbes,* September 29, 2012.

271. **Silk Road attracted a base:** Financial Action Task Force, "FATF Report: Virtual Currencies: Key Definitions and Potential AML/CFT Risks," fatf-gafi.org, June 11, 2014.

271. **18 percent of drug consumers:** M. J. Barratt, J. A. Ferris, and A. R. Winstock (2014), "Use of Silk Road, the Online Drug Marketplace in the United Kingdom, Australia and the United States," *Addiction* 109: 774–83.

271. **Due to some missteps online:** D. Leinwand Leger, "A Behind-the-Scenes Look at the Federal Agents' Digital Detective Work," usatoday.com, May 15, 2014.

271. **Being cold, overstimulated, or confused:** E. Anthes, "Outside In: It's So Loud, I Can't Hear My Budget!," psychologytoday.com, September 1, 2010.

272. **Silk Road created drug users:** S. Thielman, "Silk Road Operator Ross Ulbricht Sentenced to Life in Prison," *The Guardian,* May 29, 2015.

273. **Ulbricht pleaded for leniency:** Most of the letters submitted to the court in

support of Ross Ulbricht prior to his sentencing are available online: see documentcloud.org/documents/2086667-gov-uscourts-nysd-422824-251-2.html#document/p14/a220146.

273. **"eliminated every obstacle":** K. McCoy, "Silk Road Founder Hit with Life Imprisonment," usatoday.com, June 1, 2015.

274. **"In the late hours of Tuesday night":** C. Dewey, "You Can Take Down Pirate Bay, but You Can't Kill the Internet It Created," *The Washington Post,* December 10, 2014.

275. **the impact of peer-to-peer file-sharing:** For statistics and the music industry's views on digital piracy, see the FAQ section of the Recording Industry Association of America's website, riaa.com.

275. **"*Game of Thrones* is the most pirated show":** P. Tassi, " 'Game of Thrones' Sets Piracy World Record, but Does HBO Care?," *Forbes,* April 15, 2014.

275. **It costs HBO about $6 million:** E. Sheppard, "Here's How Much It Costs to Make a 'Game of Thrones' Episode," Mic.com, April 8, 2014. See also M. Russon, "Game of Thrones season 5 breaks piracy record with 32 million illegal downloads so far," April 22, 2015, ibtimes.co.uk.

276. **anonymity and online disinhibition:** Online disinhibition is explored in J. Suler (2004), "The Online Disinhibition Effect," *CyberPsychology and Behavior* 7(3): 321–26.

276. **If enough people are breaking a law:** S. Altschuller and R. Benbunan-Fich (2009), "Is Music Downloading the New Prohibition? What Students Reveal Through an Ethical Dilemma," *Ethics and Information Technology* 11(1): 49–56.

277. **like the fifteen-year-old boy in Sweden:** M. Humphries (2011), "15-Year-Old Facing Jail Time for Downloading 24 Movies," geek.com, August 24, 2011.

277. **A new norm has effectively been created:** S. Altschuller and R. Benbunan-Fich (2009), "Is Music Downloading the New Prohibition? What Students Reveal Through an Ethical Dilemma," *Ethics and Information Technology* 11(1): 49–56. See also T. Wingrove, A. L. Korpas, and V. Weisz (2011), "Why Were Millions of People *Not* Obeying the Law? Motivational Influences on Non-Compliance with the Law in the Case of Music Piracy," *Psychology, Crime and Law* 17(3): 261–76.

278. **A move to decriminalize piracy:** C. Green, "New Internet Piracy Warning Letters Rules Dismissed as 'Toothless,' " *The Independent,* July 23, 2014.

278. **it appears the pirates have won:** S. Hamedy, "Report: Online Piracy Remains Multi-Hundred-Million-Dollar Business," *Los Angeles Times,* May 19, 2015.

280. **"Another one got caught today":** Originally called "The Conscience of a Hacker," the manifesto was written after Loyd Blankenship was arrested and was published in an online magazine under his pseudonym "The Mentor":

Phrack 1(7), January 8, 1986. For its continued relevance today, see S. Ragan, "The Hacker's Manifesto Turns 29 Years Old," csoonline.com, January 8, 2015.

281. **criminal hackers share a set of personality traits:** D. Littlejohn Shinder and M. Cross (2008), *Scene of the Cybercrime* (Burlington, Mass.: Elsevier); and Shinder's blog, techrepublic.com. See also S. Atkinson (2015), "Psychology and the Hacker: Psychological Incident Handling," SANS Institute, sans.org, June 20, 2015.

285. **The black market has proven amazingly resilient:** "The Amazons of the Dark Net," *The Economist,* November 1, 2014.

285. **"The first thing that strikes you":** J. Bartlett, "Dark Net Drug Markets Kept Alive by Great Customer Service," *Wired,* August 21, 2014. See also M. P. Aiken and C. McMahon (2014), "The CyberPsychology of Internet Facilitated Organised Crime," in Europol, *The Internet Organised Crime Threat Assessment Report (iOCTA),* The Hague, Appendix A3: 81–87, europol .europa.eu.

285. **the number of products available on Darknets:** P. H. O'Neill, "Dark Net Markets Offer More Drugs Than Ever Before," dailydot.com, May 13, 2015.

286. **anonymous marketplaces have proven to be resilient:** K. Perry, "Dark Net Drugs Adverts 'Double in Less Than a Year,'" telegraph.co.uk, July 31, 2014. See also S. Nelson, "Buying Drugs Online Remains Easy, 2 Years After FBI Killed Silk Road," usnews.com, October 2, 2015.

287. **teens in particular have flocked online:** S. Lewis, "Protect Your Teen from Dangerous Dark Net Drugs," huffingtonpost.com, February 1, 2014.

287. **Remember the Triple A Engine:** The Triple A Engine of Internet vulnerabilities (affordability, accessibility, and anonymity) that contribute to Internet-related intimacy difficulties was identified by A. Cooper (1997), "The Internet and Sexuality: Into the New Millennium," *Journal of Sex Education and Therapy* 22: 5–6.

288. **There were roughly 187,100 drug-related deaths:** For global estimates of drug deaths and usage, see United Nations Office on Drugs and Crime (2015), *World Drug Report 2015* (United Nations, New York); unodc.org.

288. **great white sharks and serial killers:** R. A. Martin, D. K. Rossmo, and N. Hammerschlag (2009), "Hunting Patterns and Geographic Profiling of White Shark Predation," *Journal of Zoology* 279(2): 111–18. See also "Great White Sharks Can Behave Like Serial Killers, Study Finds," theguardian.com, June 22, 2009.

289. **"Criminals reveal who they are":** D. Canter (2003), *Mapping Murder: The Secrets of Geographical Profiling* (London: Virgin Books).

289. **"Individuals have different routines":** A. Madero-Hernandez and B. S. Fisher (2012), "Routine Activity Theory," in F. T. Cullen and P. Wilcox

(eds.), *The Oxford Handbook of Criminological Theory* (Oxford: Oxford University Press).

291. **The incidence of teenagers turning to cybercrime:** For the story of one teen's journey into cybercrime, see L. Kelion, "Finnish Teen Convicted of More Than 50,000 Computer Hacks," bbc.com/news, July 8, 2015.

 On teens and cybercrime generally, see "Criminals Recruiting Teens for Life of Cybercrime," recruitmentgrapevine.com, June 10, 2015.

292. **Crimebook:** S. Malik, "Teenagers Jailed for Running £16m Internet Crime Forum," *The Guardian,* March 2, 2011.

293. **a quarter of organized crime in Britain:** "Cybercrime Could Become More Lucrative Than Drugs, Police Chief Warns," telegraph.co.uk, March 1, 2015.

294. **data breach of the IRS:** E. Weise, "IRS Hacked, 100,000 Tax Accounts Breached," usatoday.com, May 26, 2015.

294. **more than 3 billion people worldwide online:** International Telecommunication Union (2015), *ICT Facts & Figures: The World in 2015* (Geneva); itu.int.

295. **The exponential growth of cybercrime:** On the exponential spread of malware in recent years, see M. Goodman (2015), *Future Crimes: Everything Is Connected, Everyone Is Vulnerable and What We Can Do About It* (New York: Doubleday), 14–15.

295. **Only 5 percent of the malware was found:** M. Garnaeva, V. Chebyshev, D. Makrushin, and A. Ivanov, "IT Threat Evolution in Q1 2015," securelist .com, May 6, 2015.

296. **we could identify, support, and nurture the most tech-talented:** On a positive note, in 2015 the U.K. government launched an initiative to find and train people with cyber skills to bolster the cyber-security industry. Through school programs and such avenues as national math competitions, Cyber First is intended to identify potential candidates for such things as undergraduate funding and government work placements: A. Stevenson, "UK Government Launches Cyber First Recruitment Drive for Future White Hats," V3.co.uk, March 25, 2015. See also the announcement of the new UK National Cyber Security Centre, "New National Cyber Security Centre set to Bring UK Expertise Together," gov.uk, March 18, 2016.

Chapter 9: The Cyber Frontier

304. **We desperately need some balance:** J. Berland (2001), "Cultural Technologies and the 'Evolution' of Technological Cultures," in Andrew Herman and Thomas Swiss (eds.), *The World Wide Web and Contemporary Cultural Theory* (New York: Routledge), chapter 12.

310. **claims that it is a "near certainty":** E. Zolfagharifard, "Stephen Hawking

Says It Is a 'Near Certainty' Technology Will Threaten Humanity Within 10,000 Years," dailymail.co.uk, January 19, 2016.

314. **Microsoft had to quickly delete:** H. Horton, "Microsoft Deletes 'Teen Girl' AI After It Became a Hitler-Loving Sex Robot Within 24 Hours," telegraph .co.uk, March 24, 2016.

INDEX

A

J

R

S

U

V

W

Y

Z

ABOUT THE AUTHOR

DR. MARY AIKEN is the world's foremost forensic cyberpsychologist. She is the director of the Cyber-Psychology Research Network and an advisor to Europol, and has conducted research and training workshops with multiple global agencies from INTERPOL to the FBI and the White House. Her research interests include cyber-security, organized cybercrime, cyberstalking, technology-facilitated human trafficking, and the rights of the child online. She is a member of the advisory board of the Hague Justice Portal, a foundation for international peace, justice, and security. Her groundbreaking work inspired the CBS television series *CSI: Cyber.* She is based in Ireland.

maryaiken.com
@maryCyPsy